Springer Series in Chemical Physics

Volume 114

The purpose of this series is to provide comprehensive up-to-date monographs in both well established disciplines and emerging research areas within the broad fields of chemical physics and physical chemistry. The books deal with both fundamental science and applications, and may have either a theoretical or an experimental emphasis. They are aimed primarily at researchers and graduate students in chemical physics and related fields.

More information about this series at http://www.springer.com/series/676

Joost Frenken · Irene Groot
Editors

Operando Research in Heterogeneous Catalysis

Springer

Editors
Joost Frenken
Huygens–Kamerlingh Onnes Laboratory, Institute of Physics
Leiden University
Leiden
The Netherlands

and

Advanced Research Center for Nanolithography
Amsterdam
The Netherlands

Irene Groot
Huygens–Kamerlingh Onnes Laboratory, Institute of Chemistry
Leiden University
Leiden
The Netherlands

and

Gorlaeus Laboratories
Leiden University
Leiden
The Netherlands

ISSN 0172-6218
Springer Series in Chemical Physics
ISBN 978-3-319-44437-6 ISBN 978-3-319-44439-0 (eBook)
DOI 10.1007/978-3-319-44439-0

Library of Congress Control Number: 2016950751

Printed on acid-free paper

This Springer imprint is published by Springer Nature
The registered company is Springer International Publishing AG
The registered company address is: Gewerbestrasse 11, 6330 Cham, Switzerland

Preface

In modern society, catalysis is a crucial technology. Approximately 90% of all chemicals and materials we use are produced via catalysis. Furthermore, catalysis impacts around one-quarter of the world's gross domestic product. A catalyst is a chemical substance that affects the rate and selectivity of a chemical reaction without being part of its end products. The primary objectives of catalysis are to enhance the reaction rate and to yield the desired products with high selectivity and stability. Well-known examples of catalytic processes are the conversion of crude oil into gasoline and the conversion of toxic automotive exhaust gases into less harmful ones.

Due to its importance to society, heterogeneous catalysis has been studied extensively over the last 100 years. Much of our current knowledge, however, has been obtained under conditions that deviate significantly from those of practical catalysis. In the traditional surface-science approach to study heterogeneous catalysis, fundamental knowledge about the behaviour of catalysts has been obtained under (ultra)high vacuum conditions and on planar model catalysts. Many useful insights have been acquired in this way, and the important research performed in this field has been acknowledged by the Nobel Prize in Chemistry for Prof. Gerhard Ertl in 2007. The reason for the discrepancy between conditions in surface science and in practical catalysis stems from the fact that most measurement techniques, suitable for obtaining accurate results at the nanoscale, cannot perform under the typical working conditions of industrial catalysis (i.e. high pressures and high temperatures). Surface-sensitive techniques are usually limited to pressures below 10^{-5} mbar. Although there are cases, where the results obtained at low pressures can be extrapolated to industrial conditions, more and more examples are encountered where this "pressure gap" is found to be accompanied by fundamental changes in the reaction mechanisms and the structures of the active phase of the catalyst. The pressure gap seems to be the rule, rather than the exception. Therefore, the investigation of catalysis under more realistic (industrial) conditions, without compromising the sensitivity to the details on the atomic and molecular scale, inescapably forms the next arena in this research field, where major breakthroughs will be forced, both scientifically and in the technological application.

To bridge the pressure gap, the last decades have seen a tremendous effort in designing new instruments and adapting existing ones to be able to investigate catalysts in situ under industrially relevant conditions (i.e. atmospheric pressures and elevated temperatures). One approach is to build a set-up in which ultrahigh vacuum chambers are combined with high-pressure chambers, using differential pumping. With this approach, X-ray photoelectron spectroscopy (XPS), low-energy ion scattering (LEIS), and transmission electron microscopy (TEM) can operate in the mbar regime. Another approach is the use of micro- or nanoreactors that separate the high-pressure volume from the (ultra)high vacuum part via ultrathin walls of an inert material. Examples are TEM and X-ray microscopy.

Some surface-science techniques are able to bridge the pressure gap without facing major limitations, e.g. scanning probe microscopy (SPM), sum frequency generation (SFG) laser spectroscopy, polarization modulation infrared reflection absorption spectroscopy (PM-IRAS), ultraviolet Raman spectroscopy, surface X-ray diffraction (SXRD), X-ray absorption spectroscopy (XAS), and ellipso-microscopy. When combined, these techniques have the possibility to determine the detailed influence of the gas environment on the structure of model catalyst surfaces, to identify active sites for catalytic processes, and to elucidate the role of promoters, all by probing surfaces with (near)-atomic sensitivity under the high-pressure, high-temperature conditions of industrial catalysis.

This book discusses the topic of *operando* research in heterogeneous catalysis. Here, the term *operando* refers to not only monitoring the catalytic surface under industrial reaction conditions, but in addition to simultaneously monitoring the reactants and products. Thereby, the relationship between the morphological and/or chemical composition of the active phase of the catalyst and its activity and selectivity can be investigated. This book covers a wide range of measurement techniques and theoretical tools that are now emerging and that are able to bridge the pressure gap. In addition to describing how they measure or compute relevant properties of catalytic reactions on solid surfaces under industrially relevant conditions of atmospheric pressures and elevated temperatures, it also presents several of the exciting new insights on heterogeneous catalysis that have been obtained recently this way.

The following topics will be discussed: the ReactorSTM and ReactorAFM (Chap. 1), Ambient-Pressure X-ray Photoelectron Spectroscopy (Chap. 2), Surface-Sensitive X-ray Diffraction Across the Pressure Gap (Chap. 3), X-ray Absorption and Emission Spectroscopy under Realistic Conditions (Chap. 4), *Operando* Transmission Electron Microscopy (Chap. 5), Planar Laser Induced Fluorescence Applied to Catalysis (Chap. 6), Ab Initio Thermodynamics and First-Principles Microkinetics for Surface Catalysis (Chap. 7), and Catalysis Engineering: From the Catalytic Material to the Catalytic Reactor (Chap. 8).

Most of all, this book illustrates that we are at the brink of a new and exciting era of surface science. Finally, after half a century of promise, the emerging new experimental and theoretical tools are bringing this research field to a more complete level of understanding and to truly predictive power in the area of catalysis. This will prove necessary to progress towards one of the holy grails of catalysis research, namely to leave the traditional path of "educated trial and error" and to enable the direct development of "designer catalysts".

Leiden, Amsterdam, The Netherlands Joost Frenken
Leiden, The Netherlands Irene Groot

Contents

Contributors

Sara Blomberg Division of Synchrotron Radiation Research, Lund University, Lund, Sweden

Hendrik Bluhm Lawrence Berkeley National Laboratory, Chemical Sciences Division, Berkeley, CA, USA

Jeroen A. van Bokhoven Institute for Chemical and Bioengineering, ETH Zurich, Zurich, Switzerland; Paul Scherrer Institute, Villigen, Switzerland

Mauro Bracconi Laboratory of Catalysis and Catalytic Processes - Dipartimento di Energia, Politecnico di Milano, Milan, Italy

Alberto Cuoci Dipartimento di Chimica, Materiali e Ingegneria Chimica "G. Natta", Politecnico di Milano, Milan, Italy

Joost Frenken Huygens–Kamerlingh Onnes Laboratory, Leiden University, Leiden, The Netherlands; Advanced Research Center for Nanolithography, Amsterdam, The Netherlands

Irene Groot Huygens–Kamerlingh Onnes Laboratory, Leiden University, Leiden, The Netherlands; Gorlaeus Laboratories, Leiden University, Leiden, The Netherlands

Johan Gustafson Division of Synchrotron Radiation Research, Lund University, Lund, Sweden; Lund University, Lund, Sweden

Urs Hartfelder Institute for Chemical and Bioengineering, ETH Zurich, Zurich, Switzerland

Osman Karslıoğlu Lawrence Berkeley National Laboratory, Chemical Sciences Division, Berkeley, CA, USA

Patricia Jane Kooyman Chem Eng, University of Cape Town, Rondebosch, South Africa

Edvin Lundgren Lund University, Lund, Sweden; Division of Synchrotron Radiation Research, Lund University, Lund, Sweden

Matteo Maestri Laboratory of Catalysis and Catalytic Processes - Dipartimento di Energia, Politecnico di Milano, Milan, Italy

Maarten Nachtegaal Paul Scherrer Institute, Villigen, Switzerland

Stefano Rebughini Laboratory of Catalysis and Catalytic Processes - Dipartimento di Energia, Politecnico di Milano, Milan, Italy

Karsten Reuter Chair for Theoretical Chemistry and Catalysis Research Center, Technical University Munich, Garching, Germany

Andreas Stierle Deutsches Elektronen Synchrotron DESY, Hamburg, Germany

Johan Zetterberg Division of Combustion Physics, Lund University, Lund, Sweden

Jianfeng Zhou Division of Combustion Physics, Lund University, Lund, Sweden

Chapter 1
Live Observations of Catalysts Using High-Pressure Scanning Probe Microscopy

Joost Frenken and Irene Groot

Abstract Recently it has become clear that essential discrepancies exist between the behavior of catalysts under industrial conditions and the (ultra)high vacuum conditions of traditional laboratory experiments. Differences in structure, composition, reaction mechanism, activity, and selectivity have been observed. These differences indicated the presence of the "pressure gap", and made it clear that meaningful results can only be obtained at high pressures and temperatures. This chapter focuses on the development of scanning probe microscopy for *operando* observations of active model catalysts. We have developed instrumentation that combines an ultrahigh vacuum environment for model catalyst preparation and characterization with a high-pressure flow reactor cell, integrated with either a scanning tunneling microscope or an atomic force microscope. We combine the structural observations obtained under high-pressure, high-temperature conditions with time-resolved mass spectrometry measurements on the gas mixture leaving the reactor. In this way, we can correlate structural changes of the catalyst due to the gas composition with its catalytic performance. This chapter provides an overview of the instruments we developed and illustrates their performance with results obtained for different model catalysts and reactions.

J. Frenken · I. Groot (✉)
Huygens–Kamerlingh Onnes Laboratory, Leiden University,
P.O. Box 9504, 2300 RA Leiden, The Netherlands
e-mail: i.m.n.groot@lic.leidenuniv.nl

J. Frenken
e-mail: j.frenken@arcnl.nl

J. Frenken
Advanced Research Center for Nanolithography,
P.O. Box 93019, 1090 BA Amsterdam, The Netherlands

I. Groot
Gorlaeus Laboratories, Leiden University,
P.O. Box 9502, 2300 RA Leiden, The Netherlands

J. Frenken and I. Groot (eds.), Operando *Research in Heterogeneous Catalysis*,
Springer Series in Chemical Physics 114, DOI 10.1007/978-3-319-44439-0_1

1.1 Introduction

The topic of heterogeneous catalysis has been studied extensively in the last one hundred years. Much of our current knowledge has been obtained under conditions that differ significantly from those of practical catalysis. The reason for this discrepancy stems from the fact that most techniques, suitable for obtaining accurate results at the nano-scale, cannot perform under the typical working conditions of industrial catalysis (i.e. high pressures and high temperatures). Surface-sensitive techniques, such as Auger electron spectroscopy (AES), X-ray photoelectron spectroscopy (XPS), or low-energy electron diffraction (LEED), are limited to pressures below 10^{-5} mbar [1]. Although there are cases, where the results obtained at low pressures can be extrapolated to industrial conditions [2, 3], more and more examples exist where this "pressure gap" is found to fundamentally change the reaction mechanisms [4–9]. The pressure gap seems to be the rule, rather than the exception. The investigation of catalysis under more realistic (industrial) conditions, without compromising the sensitivity to the details on the atomic and molecular scale, inescapably forms the next arena in this research field, where major breakthroughs will be forced, both scientifically and in the technological application. But the difference between laboratory studies and chemical industry in terms of pressure and temperature is not the only gap in heterogeneous catalysis research. Also the nature of the catalyst used is very different. In academic studies the catalyst is often simplified to a model system, usually consisting of a single-crystal metal surface, as opposed to industrial catalysts, which generally consist of metal nanoparticles deposited on oxidic, nanoporous supports and are accompanied by additional elements such as promoters, fillers, and binders. This discrepancy is commonly known as the "materials gap".

1.1.1 Why Working Conditions?

Recently it was discovered that reaction mechanisms taking place under typical laboratory conditions of (ultra)high vacuum (10^{-5}–10^{-11} mbar) can differ significantly from those taking place under high-pressure (HP) conditions (mbar–bar range) [4–9]. Therefore, the last decades have seen a tremendous effort in adapting existing surface-science techniques to high-pressure conditions. One approach is to build a set-up in which ultrahigh vacuum chambers are combined with high-pressure chambers, using differential pumping. With this approach, X-ray photoelectron spectroscopy [10, 11], low-energy ion scattering (LEIS) [12], and transmission electron microscopy (TEM) [13, 14] can operate in the mbar regime. Another approach is the use of micro- or nano-reactors that separate the high-pressure volume from the (ultra)high vacuum part via ultrathin walls of an inert material. Examples are TEM [15] and X-ray microscopy [16].

Some surface-science techniques are able to bridge the pressure gap without facing major limitations, e.g. scanning probe microscopy (SPM) [17–21], sum frequency generation (SFG) laser spectroscopy [22–24], polarization-modulation infrared reflection absorption spectroscopy (PM-IRAS) [25], ultraviolet Raman spectroscopy [26–28], and ellipso-microscopy [29]. Furthermore, the successful application of surface X-ray diffraction (SXRD) [30–32] and X-ray absorption spectroscopy [33, 34] to study catalytic surfaces at high temperatures and pressures was recently published.

Scanning probe microscopy (e.g. scanning tunneling microscopy (STM) and atomic force microscopy (AFM)) is one of the few atomically-sensitive techniques that in principle can operate under conditions varying from millikelvins to temperatures as high as 1000 K [35–37], and pressures varying from 10^{-11} mbar to above 1 bar. These techniques have the possibility to determine the detailed influence of the gas environment on the structure of model catalyst surfaces, to identify active sites for catalytic processes, and to elucidate the role of promoters, all by imaging surfaces with (near)-atomic resolution under the relevant high-pressure, high-temperature conditions of industrial catalysis. There are several practical issues that can affect the quality of the images and the resolution. In addition, when aiming to detect reaction products simultaneously, further complications may arise.

This chapter is organized as follows. We start with a general introduction about STM, followed by a section about the general requirements for the development of SPM instruments that are dedicated to catalysis research under relevant industrial conditions. The chapter concludes with examples of catalytic systems studied using *operando* STM, namely CO oxidation, NO reduction, Fischer-Tropsch synthesis, hydrodesulfurization, and chlorine production.

1.2 Scanning Tunneling Microscopy

The scanning tunneling microscope (STM) was invented in 1982 at IBM Zürich by Gerd Binnig and Heinrich Rohrer [38, 39] and implemented by Gerd Binnig, Heinrich Rohrer, Christoph Gerber, and Erich Weibel [40, 41]. The STM was an immediate success, and Binnig and Rohrer were awarded the Noble Prize in Physics in 1986 for their invention. Nowadays, the STM is an indispensable instrument in a physics laboratory, and it is also employed regularly in surface-science and catalysis research.

The main components of an STM consist of [35, 42]:

- An atomically sharp tip. STM tips are usually made from W, Au, or PtIr wire. They are fabricated by cutting or etching the metal wire. *In situ* treatments consist of annealing and sputtering.
- A scanner to raster the tip over the sample surface. This is done by using piezoelectric ceramics, as they can convert electric signals of 1 mV to 1 kV into mechanical motion in the range of fractions of an Å to a few μm.

- A coarse positioning system to choose the region that is to be imaged, to bring the tip within tunneling distance of the sample, and to retract it a few mm when e.g. placing and removing the sample in/from the STM.
- Vibration isolation. For stable operation of the STM, the influence of external vibrations on the tip-sample distance should be minimized, e.g. to 0.01 Å or less. Vibration isolation is usually achieved by suspending the inner STM stage, including tip and sample, by springs, often in combination with a damping system, for example employing eddy currents. In addition, STM setups are often isolated as a whole from external vibrations, for example by use of passive or active isolator legs.
- Feedback electronics to control the tip-sample distance.
- A computer system to control the tip position, to acquire data, and to convert the data into an STM image.

The working principle of an STM is based on the quantum mechanical phenomenon of tunneling: the penetration of electrons through a potential barrier. When the sharp tip of the microscope is placed close enough to the sample surface (5–10 Å), the wave function(s) of the atom(s) at the end of the tip will overlap with the wave functions of the atoms in the sample surface. When a bias voltage is applied between the tip and sample, a tunneling current will flow through the tip-sample gap. Figure 1.1 shows a schematic diagram of the electron energies on the side of the sample and those on the side of the tip, involved in the tunneling process. The tunneling current is given by the following equation:

$$I(d) \propto eVexp(-\kappa d), \tag{1.1}$$

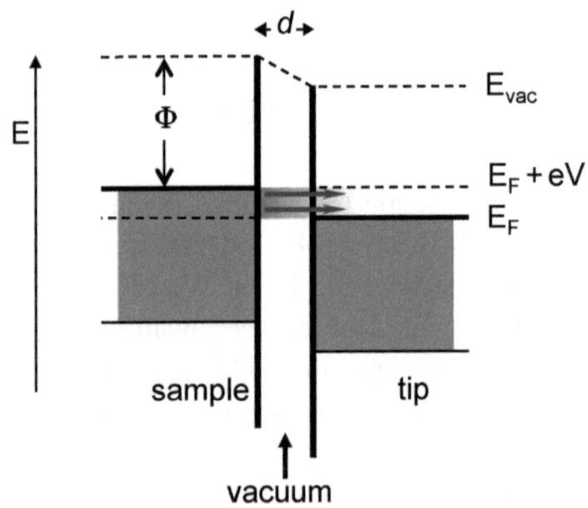

Fig. 1.1 Schematic energy diagram for the sample and the tip in an STM experiment. d is the tip-sample distance, Φ the work function, which we have chosen identical here for sample and tip, E_{vac} is the vacuum level and E_F the Fermi level. The applied bias voltage V offsets the Fermi levels of the sample and the tip by an energy eV and gives rise to a net tunneling current between the two, proportional to eV and depending exponentially on the distance d and the height of the tunneling barrier Φ

where $I(d)$ is the tunneling current as a function of the tunneling gap, V is the applied bias voltage, $\kappa \propto \sqrt{\Phi}$, where Φ is the (average) work function of the sample and the tip, and serves as the effective tunneling barrier height, and d is the tunneling gap. This description is an approximation, only valid at very low tunneling currents, and in the case of equal work functions for tip and sample [35, 42]. The sharp dependence of the tunneling current on the tip-sample distance determines the extremely high vertical resolution of STM. Typically a change of tip-sample distance of 1 Å results in a change in the tunneling current of nearly an order of magnitude. The lateral resolution is determined by the fact that up to 90% of the tunneling current flows between the "last" atom of the tip and the surface atom closest to it. When scanning the tip along the surface, the local density of states of the surface atoms is probed. When the applied sample bias voltage is negative with respect to the tip, the filled states are probed, and when this applied voltage is positive, the empty states are probed.

Three main modes of STM operation are defined, depending on the variation of the lateral coordinates x and y, the height z, the bias voltage V, and the tunneling current I. In the constant current mode, I and V are kept constant, x and y are varied by scanning the tip over the surface, and z is measured. This is the most widely used scanning mode. To maintain constant tunneling current, the feedback electronics constantly adjust the vertical position of the tip by varying the feedback voltage on the z-piezoelectric element. For an electronically homogeneous surface, this essentially means that the tip-sample distance is kept constant. The height variations of the surface are inferred from the variations in the feedback voltage as a function of lateral position (x, y). This mode can be used even for very rough or strongly inclined surfaces. The scan speed, however, is limited by the finite response time of the feedback electronics. In the constant height mode, z and V are kept constant, x and y are again varied by scanning the tip over the surface, and I is measured as a function of the tip position. In this mode, the feedback electronics is turned off or reduced to a low-frequency response, only to follow the slope and gentle height variations. The higher-frequency signature of the atomic-scale structure is not followed by the tip and is recorded as current variations. Since feedback is not required at this high frequency, this type of scanning can be performed at high speed. However, this mode can only be applied for relatively flat surfaces. In the various scanning tunneling spectroscopy (STS) modes, the voltage V is varied. Since the tunneling current is determined by summing over electron states in the energy interval determined by the bias voltage, varying this voltage will give information on the local density of states as a function of energy. When probing the electronic properties of a very local area or even a single atom, chemical information about the surface species can be obtained. For an extensive overview of the working principles of STM, see e.g. [35, 42].

1.3 High-Pressure SPM Instrumentation

The development of STMs operating at (near)-ambient reaction conditions started as early as 1993, with the pioneering work of McIntyre and co-workers [43]. In their design, the STM is contained in a high-pressure cell, consisting of a 2 L stainless-steel vacuum chamber. Using an infrared spot heater, the sample can be heated to 1400 K under atmospheric pressures. In addition, the system is equipped with a differentially-pumped quadrupole mass spectrometer (QMS), which can be used for gas analysis. A few years after this introduction the design was combined with a UHV chamber. Instead of a QMS, a gas chromatograph (GC) was used for gas analysis [44].

After this successful introduction of high-pressure STM, several other designs have been introduced [19, 45–48]. All these HP-STMs had in common that they include a microscope inside a high-pressure reactor, that is connected to a UHV chamber for sample preparation and characterization using common surface-science techniques such as AES, XPS, and LEED. In our group we have adopted a fundamentally different design concept, the ReactorSTM, in which not the entire microscope, but only the tip and its holder, are contained in the high-pressure environment [17]. The motivation for this radically different design is as follows. To be able to fully explore the effect of surface restructuring of the catalyst on its activity, the reaction rate needs to be measured simultaneously with structural measurements using STM. This can be done by operating the cell that contains the STM as a flow reactor, via analysis of the gas leaving the reactor using a QMS or GC. However, when using a cell large enough to contain the entire microscope, the low conversion rates that should be expected on the small active areas of typical model catalysts make that one can obtain measurable product concentrations only at very low flow rates, which would introduce extremely long response times. Furthermore, large reactors also introduce high probabilities for reactants and reaction products to adsorb on and react with the walls of the reactor. These risks also extend to the components of the STM itself, such as the piezo element. The latest version of the ReactorSTM setup is a versatile instrument that provides atomic resolution under harsh reaction conditions [20].

In addition to high-pressure STMs, some high-pressure AFMs have been developed as well. Again, the simplest approach is to operate a standard AFM in a high-pressure cell [49, 50]. However, this severely limits the operating temperature range and the use of corrosive gases. A more advanced approach separates the high-pressure flow cell from the piezo of the AFM scanner with a membrane. In this way, the AFM can operate up to 423 K and 6 bar in liquid [51], or up to 350 K and 100 bar in supercritical CO_2 [52]. These instruments are limited to contact-mode AFM and, due to long equilibration times, to constant temperature. To be able to perform real catalytic studies, we have developed an AFM [21] based on the same principles as the ReactorSTM [20]. We continue in the next sections with further details on the design and performance of these special STM and AFM set-ups.

1.3.1 *Requirements for* Operando *SPM*

As mentioned in the previous section, an unconventional design for SPM tools suitable for catalytic studies under reaction conditions has been developed in our group. For the development of both the ReactorSTM [20] and the ReactorAFM [21] the following requirements were taken into account:

- The microscope must be suitable for catalytic studies under realistic conditions, i.e. high pressures (1 bar and beyond) and elevated temperatures (at least 400 K). Under these conditions, the microscope must be capable of imaging the catalytic surface with atomic resolution (STM) or step resolution (AFM) and low drift.
- The set-up must be suitable for *operando* studies, i.e. studies combining the structural information of the catalyst with its activity. For this, the time resolution for measurements of the reaction product needs to be 10 s or better.
- The design of the instrument should prevent chemistry taking place anywhere else than on the catalyst surface. Reactions taking place on the reactor walls or elsewhere will not only hamper the accuracy of activity measurements, but may also corrode components of the microscope.
- Measurements should take place on well-defined model catalysts that are initially atomically clean and highly-ordered. The catalyst surface must not be exposed to air between preparation and characterization under reaction conditions.

As explained in Sect. 1.1.1, an STM or AFM has no intrinsic difficulties working at atmospheric pressures and beyond. Due to the small tip-surface distance no significant interference by the surrounding gas molecules is expected. Effects of elevated temperatures include an increase in thermal drift. Using special materials with low expansion coefficients, these effects can be minimized and stable imaging at temperatures as high as 1300 K can be achieved [53]. However, when combining high temperatures with high pressures difficulties may arise. The piezo element, that is used to actuate the tip, must be kept under its Curie temperature. In vacuum conditions, large temperature differences can be maintained over short distances. In the presence of a gas atmosphere, however, heat transport via the gas renders this challenging. In addition, a large temperature difference between sample and piezo causes a convective flow in the gas atmosphere, resulting in erratic drift in the images. We have circumvented this problem by keeping most of the components of the microscope in vacuum, while the sample is exposed to reaction conditions.

To be able to perform catalytic studies under *operando* conditions, the catalytic surface must be contained inside a reactor volume, that, in addition to housing the microscope, is connected to a means of measuring catalytic activity (either via QMS or GC). To optimize both the time resolution and the sensitivity to changes in the concentrations of reactants and products, the ratio between the surface area of the model catalyst and the volume of the reactor, including the gas line connecting the reactor to the gas analysis system, must be maximized. Since SPM can only be used on flat surfaces or on clusters of nanoparticles on flat supports, we are limited to very small surface areas. To obtain a high ratio between the active surface area and the reactor volume, the reactor volume must be as small as possible.

Under high-pressure, high-temperature conditions, chemical reactions can take place not only on the surface of the model catalyst, but also on the sides and back face of the sample. Furthermore, chemical reactions can take place on other surfaces exposed to the reactive gases such as the walls of the reactor, the sample holder, and elements of the microscope. These undesired reactions not only influence the outcome of the experiment (e.g. higher reactivity can be observed than if the only reactivity measured would be due to the catalyst), it can also seriously damage the instrument, especially when aggressive reactions are studied. To eliminate these undesired reactions as much as possible, inert materials are chosen for the reactor walls. In addition, we keep as much as possible the elements of the microscope and the side and rear faces of the sample outside the reactor volume.

To prepare well-defined, highly-ordered model catalysts, standard cleaning and preparation procedures are employed. These methods, such as ion sputtering, annealing, physical vapor deposition of metals, and gas exposure, require UHV conditions. Techniques to investigate crystal composition, structure, and ordering, such as LEED, AES, and XPS, can only be employed under UHV conditions. Since the prepared catalyst samples cannot be exposed to air when transferring them to the high-pressure reactor and SPM, a combination of a UHV chamber with the microscope and reactor is necessary.

1.3.2 Design of the ReactorSTM

Figure 1.2 shows a schematic drawing of the concept of the ReactorSTM, in which all of the above-mentioned design criteria are met [17, 20]. The ReactorSTM integrates a very small flow reactor (0.5 mL) having inert walls with a scanning tunneling microscope. The reactor volume is placed inside a UHV system, used for the preparation and characterization of model catalysts. Two gas lines, one for the inlet of gases, one for the outlet, are connected to the reactor. The inlet line is connected to a gas system which controls the flow, mixing ratio, and pressure of the reactant gases (the gas system will be described in more detail in Sect. 1.3.4). The outlet line is connected to the back pressure controller in the gas system, after it has passed a small chamber housing the QMS for analysis of the reactant and product gases that leave the reactor.The only parts of the microscope that are inside the reactor volume are the tip and tip holder. The reactor is sealed off from the surrounding UHV using two flexible O-rings. The lower O-ring (Viton) separates the reactor from the piezo element, that is used to actuate the motion of the tip. The upper O-ring, consisting of Kalrez (Dupont), a chemically resistant material, is pressed between the sample and upper part of the reactor body, thus sealing off the high pressures inside the reactor from the surrounding UHV environment. The sample can be radiatively heated from the rear (the upper side in Fig. 1.2).

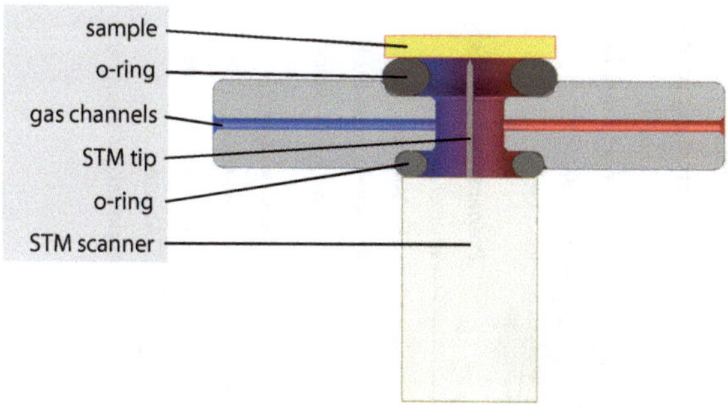

Fig. 1.2 Conceptual drawing of the ReactorSTM. The STM tip is contained in a small high-pressure volume, while the STM scanner is not exposed to the gases. The sample forms one side of the reactor, while the other walls are chemically inert. Two polymer O-rings seal off the high-pressure volume from the UHV system around it. This figure was reprinted from [20] with permission from the American Institute of Physics

Figures 1.3 and 1.4 show the configuration of the UHV system. It consists of three chambers separated by gate valves. The first chamber is the XPS chamber. Here, the XPS for use under UHV conditions is located. Furthermore, this chamber houses a sample library, and a sample load-lock ensures the loading of samples without breaking the vacuum. The second chamber is the preparation chamber. It contains an ion sputtering gun, a combined LEED/AES system, and an electron-beam evaporator for preparation of metallic nanoparticles. In addition, gas lines with leak valves are present for sample cleaning and catalyst preparation. The third chamber is the STM chamber, which houses the microscope integrated with a small flow reactor. On the top flange of this chamber a seal library has been installed for the Kalrez seals that close off the reactor. Using a wobble stick seals can be placed on and removed from the top plate of the reactor without breaking the vacuum. Sample transfer between the chambers is performed with a rack-and-pinion transfer rod.

In Fig. 1.5 a schematic cross section of the sample holder and the reactor with the microscope is shown. The upper seal consists of a custom-made Kalrez ring (Dupont) that is vulcanized onto a stainless-steel holder, enabling exchange of the seals without breaking the vacuum. Due to the specifications of the Kalrez seal, the operation temperature is limited to 600 K. The STM body is fabricated from Zerodur (Schott), a chemically resistant glass type with low thermal expansion coefficient, minimizing thermal drifting of the STM during temperature changes.

The sample holder is made out of Invar, a low-expansion steel, again to minimize thermal drift. It is strongly pressed against the top of the STM body, making hard mechanical contact via three adjustable screws. In this way, a short and stiff mechanical loop is established between the sample and the tip, which is essential for high-quality imaging. When pressing the sample holder against the STM body, the Kalrez seal is compressed to 80% of its original thickness, providing a leaktight seal. When the reactor is exposed to pressures up to 6 bar, UHV conditions are maintained

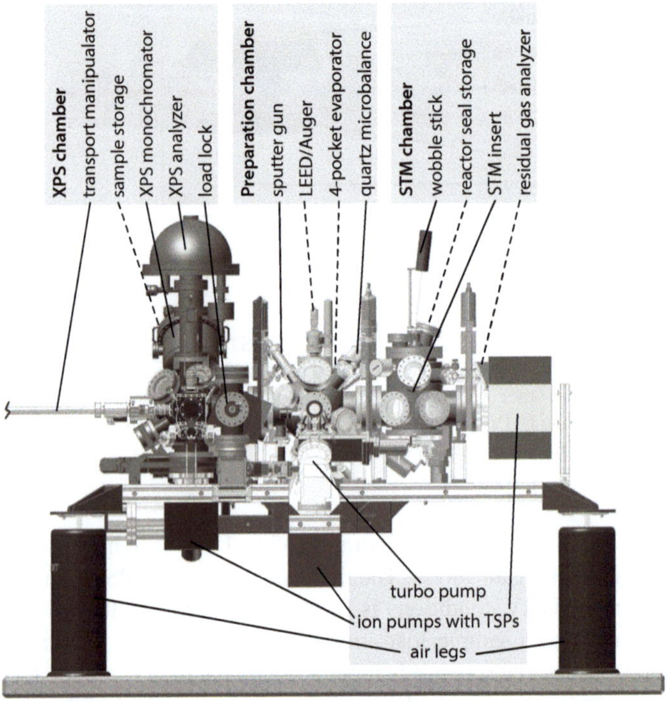

Fig. 1.3 Schematic drawing of the UHV system showing the three chambers with the equipment for sample preparation and characterization. *Dashed lines* point to components that are not visible. This figure was reprinted from [20] with permission from the American Institute of Physics

Fig. 1.4 Photograph of the UHV system showing the chambers with the equipment for sample preparation and characterization. This version does not have the XPS-system incorporated. Photograph courtesy of Leiden Probe Microscopy B.V.

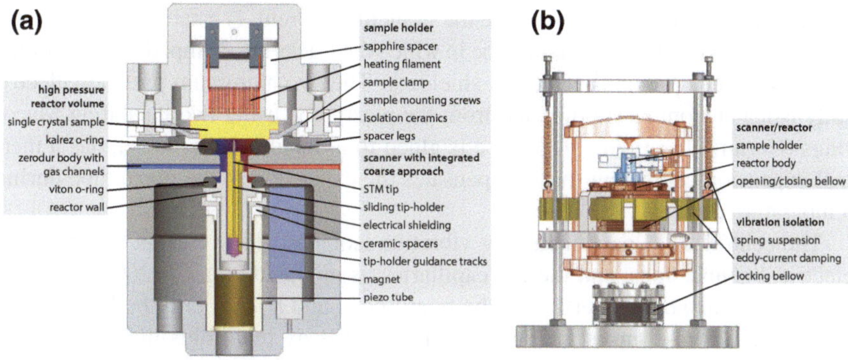

Fig. 1.5 Detailed schematic of **a** the sample holder, the reactor and the scanner (viewed in cross section), and of **b** the complete STM insert, including its vibration isolation and damping, mounted on a CF-200 flange. This figure was reprinted from [20] with permission from the American Institute of Physics

in its surroundings. The current limitation of 6 bar is due to the gas system, not to the seal. Figure 1.6 shows a photograph of the ReactorSTM assembly with the reactor opened and the sample holder in place.

A single piezo tube (EBL2, EBL Products) is used for both the coarse approach and the fine scanning motion of the tip. The STM tip is clamped in a steel holder, which is pulled against two steel tracks using a SmCo magnet (IBS Magnet). The tip holder and tracks are Au-coated, to ensure chemical inertness and to optimize the stick-slip behavior of the motor. The force of the magnet is determined by its size, shape, and position. It is optimized such, that the maximum acceleration that can be generated along the length axis of the piezo tube is high enough to overcome the static friction force between the tip holder and the tracks. The electrical connec-

Fig. 1.6 Photograph of the open reactor with the sample holder in place

tion to the tip, necessary to measure the tunneling current, is established via the tip holder, tracks, and the aluminum tube in which the tracks are clamped. An additional aluminum piece provides electrical shielding. The two aluminum parts are electrically isolated from each other and from the piezo tube using two insulating Macor rings (Corning Inc.). The piezo tube is glued to a titanium base, which has a thermal expansion coefficient that compensates the expansion of the piezo tube during temperature changes.

The central STM part, including vibration isolation, eddy current damping, all electrical connections, and the gas capillaries, is mounted on a CF-200 flange (see Fig. 1.5b). Thus, easy insertion of the microscope and reactor into the SPM chamber of the UHV set-up is guaranteed. The STM body that holds the microscope, the reactor, and the sample holder is suspended by a set of springs combined with an eddy current damping system, in order to isolate the STM from external, mechanical vibrations. When transferring samples, the STM body can be locked via controlled inflation of a bellow. A second bellow can be inflated via a thin capillary to press the reactor body to the Kalrez seal, thus closing off the reactor volume from the UHV surroundings. The STM is controlled using fast analog/digital SPM control electronics (Leiden Probe Microscopy B.V.) capable of video-rate STM imaging [54].

1.3.3 Design of the ReactorAFM

The instrument described in the previous section, the ReactorSTM, is able to bridge the pressure gap. Scanning tunneling microscopy, however, can only be performed on conductive samples, usually metal single crystals or thin oxide films deposited on metal surfaces. To study metallic nanoparticles deposited onto porous oxide supports, and thereby also bridging the so-called materials gap, a different scanning probe technique is required: the atomic force microscope. While STM uses an electrical current to probe the sample, AFM uses the interaction force between tip and sample, and is therefore independent of its conductivity. To be able to investigate more realistic catalysts, we have developed the ReactorAFM [21]. Its design is based on the proven concept of the ReactorSTM (see Sect. 1.3.2), but its capability to image supported nanoparticles adds unique value for *operando* catalysis research.

The ReactorAFM must be able to resolve nanoparticles supported on flat surfaces under industrially relevant conditions with sufficient detail. The minimum requirements that we have defined are a lateral resolution of 1 nm, and a vertical resolution of 0.1 nm, with a range of at least 1 μm in each direction. Furthermore, the scanner should be sufficiently stable to enable imaging of a single feature on the surface for at least one hour, thus placing constraints on the thermal drift of both the scanner and force sensor. Additional requirements are discussed in Sect. 1.3.1, and most design specifications chosen to meet these requirements are described in Sect. 1.3.2. The AFM scanner, however, differs significantly from the STM scanner. The AFM scanner is based on the piezoelectric read-out of a quartz tuning fork (QTF). Due to the small volume of the reactor, no optical access to the tip is possible, ruling out laser deflection techniques commonly used in AFM. Since quartz is chemically inert, and exceptionally high resolution has been reported using QTF-based AFM [55, 56], this is the best choice for the ReactorAFM. The high stiffness makes the QTF also relatively insensitive to the presence of the gas atmosphere, thus enabling us in principle to extend imaging with good resolution to reaction conditions.

The QTF with the AFM tip is mounted on a rod that is magnetically clamped inside the piezo tube. Using a stick-slip motion, the rod can slide up and down during the coarse approach of the tip to the sample. The rod consists of two halves and is held against two tracks using a SmCo magnet. The tracks supporting the tip holder are mounted on a capped polyetherimide (PEI) cylinder, located inside the piezo tube. An additional cylinder, made of aluminum, between the PEI component and the piezo tube provides electrical insulation from the high piezo voltages. The PEI cylinder also forms part of the reactor walls, so the piezo tube is not exposed to high-pressure gases. This avoids chemical and thermal stability issues. The two tracks are also used as feedthroughs for the two electrical signals of the QTF through the PEI reactor wall. Each track is in contact with one of the two halves of the tip holder. The tracks traverse the insulating PEI component and are connected by coaxial cables to floating-shield BNC feedthroughs on the CF-200 flange of the insert.

The QTF used is a commercial miniature crystal with a resonance frequency of 32.768 kHz (Micro Crystal AG). The tuning fork is modified to obtain the required overall dimensions of $1.9 \times 0.5 \times 0.12\,\text{mm}^3$, and the fundamental resonance frequency is ~96 kHz. The QTF is mounted on the tip holder in the QPlus configuration [57], i.e. the lower prong is completely fixed and the upper prong acts as a single piezoelectric cantilever. After fixating the lower prong, the Q-factor of the first resonance at ambient conditions is 3×10^3. A ceramic spacer is used to tilt the QTF to an angle of 2.5° to ensure that the apex of the upper prong is the first part to come into contact with the sample surface. Figures 1.7 and 1.8 show the resulting assembly.

A micrometer-sized AFM tip is grown at the apex of the upper prong of the QTF, using electron-beam induced deposition (EBID) of platinum employing a scanning electron microscope (SEM). After preparation, the tip consists of 16 atom % platinum, the remainder being amorphous carbon [58]. The electrical connections from the tuning fork electrodes to the tip holder are made by ball bonding using 25 μm diameter Au wire. The electrical path continues via the tracks that support the tip holder, followed by coaxial cables to the UHV feedthroughs.

The ReactorAFM operates in non-contact or frequency-modulation (FM) mode. In this mode, the cantilever is oscillated at resonance with an amplitude in the range of 10 pm to 100 nm. When the tip is near the surface, the tip-sample interaction force gradient will influence the effective spring coefficient of the mechanical oscillator, resulting in a shift of the resonance frequency. In the case of dissipative forces, there is also a decrease of the amplitude and an additional phase shift. The resonance frequency is measured by a phase-locked loop. The output signal of this loop is used as the input for the height feedback loop of the AFM scanner in order to trace the surface at constant frequency shift. A separate feedback system adjusts the drive amplitude to keep the oscillation amplitude constant, thereby ensuring that the surface of con-

Fig. 1.7 Three views of the sliding rod with the QTF and the two tracks. The rod consists of two halves, each in contact with one of the two tracks. The slider has a special shape with a groove and a flat side to have a well-defined orientation and thus a well-defined combination of contacts with the two tracks. This figure was reprinted from [21] with permission from the American Institute of Physics

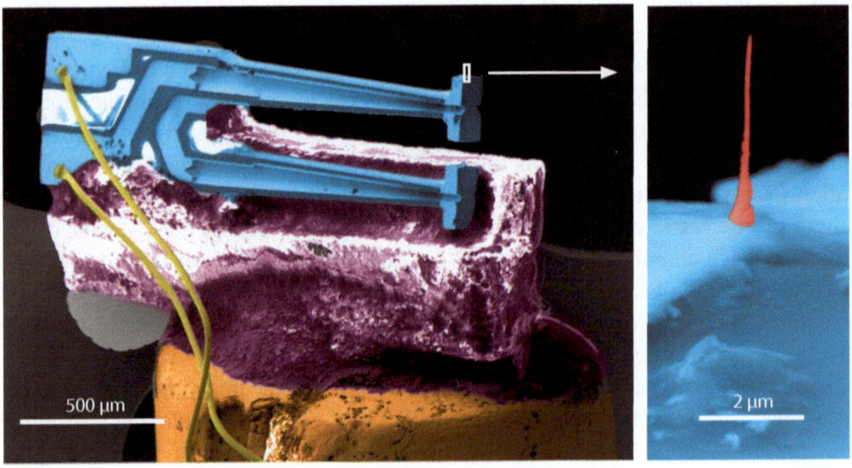

Fig. 1.8 Scanning electron microscopy images with false colors for enhanced contrast of the QTF glued on a ceramic spacer which is glued to the slider, with a close-up of the apex of the upper prong and the EBID-grown AFM tip. This figure was reprinted from [21] with permission from the American Institute of Physics

stant frequency shift corresponds to a surface of constant force gradient. The drive signal of this amplitude feedback loop is recorded in a separate channel and can be used to derive the dissipative force.

The motion of the QTF is controlled via an excitation/detection circuit located directly outside the UHV system. It is based on a circuit developed by Grober et al., which compensates for the stray capacitance of the QTF, and measures the oscillating current through the QTF with an I-V convertor when it is driven at resonance by an external oscillator voltage source [59]. The shift in resonance frequency of the QTF is detected using a Zurich Instruments HF2LI lock-in amplifier with phase-locked loop. It also supplies the oscillating drive voltage at resonance. The height feedback and scanning are performed using high-speed SPM electronics from Leiden Probe Microscopy [54].

1.3.4 Design of the Gas System

To investigate chemical reactions a gas system is required that can mix various gases over wide ranges in composition, with separate control over the gas flow rate of the individual components, and the total pressure in the mixture, and with a response time in the order of a few seconds. This latter requirement necessitates a configuration with small total volume. Furthermore, the system should enable the user to change partial pressures, flow rates, and total pressure without interrupting the imaging. This implies that pressure overshoots, which can cause tip crashes, cannot be tolerated. Finally, one has to recognize that high purity of the gases entering the reactor is essential and that dead volumes or badly refreshed components should be avoided. To separately control the gas composition, flow rate, and total pressure inside the reactor, a combination of mass flow controllers and back pressure controllers (Bronkhorst Hi-Tech) is employed. To choose different pathways for the gases through the gas system, custom-modified versions of GC-valves (Valco) are used. The advantage of these modified valves is that their design allows for complex flow patterns while still keeping the total volume very small and having no dead volume. Figure 1.9 shows an overview of the architecture of the gas manifold as designed by LPM [60]. With this system, up to five different gases can be mixed. The architecture of the rotating valve V1 is such that a gas is either flowing to rotating valve V2 or into the drain/pulse line. In this way, a certain gas flow at a certain pressure set by pressure controller BPC-pulse can be prepared in the drain/pulse line, prior to adding it to the flow through the reactor. In this way, gases can be added to the reactor flow during STM scanning, without any danger for tip crashes. Via RV3, a certain volume of a gas can be dosed to the reactor for gas-titration experiments. To enhance the dynamic range of partial pressures to be set, an additional line with a shunt mass flow controller is installed. To optimize response time, the inner diameter of the gas lines and bores of the valves are chosen to be small, but not too small to prevent unwanted large pressure drops in the system which might compromise the reactor pressure reading. The lines running towards the reactor typically have an inner diameter of 0.5 mm, the lines downstream of the reactor have an inner diameter of 0.75 mm. The gas flow controllers have a capacity up to 10 mL/min (30 mL/min for the carrier gas). Given the volume of the reactor to be 0.5 mL, the refresh rate of the reactor is up to 20 times per minute for a single gas. The response time of the system is determined by the internal volume of the gas manifold and the volume of the gas line connecting the gas manifold to the reactor. Typically it takes 10 seconds after setting up a new gas composition before it ends up at the sample in the reac-

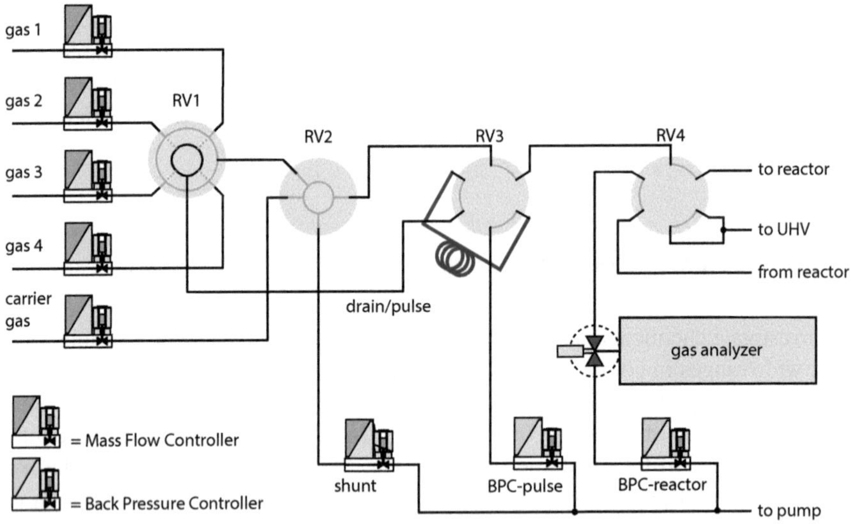

Fig. 1.9 Manifold for gas mixing and analysis. Up to four reactive gases and a carrier gas can be mixed via a computer-controlled manifold, consisting of rotating valves (RV 1–4), several mass flow controllers, and two back pressure controllers (BPC). A continuous sampling gas analyzer provides high time-resolution gas analysis. This figure was reprinted from [20] with permission from the American Institute of Physics

tor. For the gas manifold used with corrosive gases special materials are employed that can withstand these aggressive conditions (e.g. Hastelloy C gas lines and special polymer seals inside the GC-valves). The gas composition of the outgoing mixture is analyzed using a QMS. All valves of the gas system are operated by a computer. Home-written Labview and Python programs continuously log all valve settings and read-outs of the mass flow and back pressure controllers.

1.4 Catalytic Systems Investigated Under High-Pressure Conditions

In this section we discuss some catalytic systems that are currently under investigation in our group. We start out with a short summary of effects observed for CO oxidation on Pt(110) and Pd(100). With our ReactorSTM and -AFM setups we have now been able to move beyond this regime of "baby" catalysis toward more relevant reactions and have grown up to study more relevant reactions, such as NO reduction and Fischer-Tropsch synthesis. We complete the overview with first results obtained for two very aggressive systems: the hydrodesulfurization (HDS) reaction and the chlorine production reaction (Deacon process).

1.4.1 CO Oxidation on Pt(110)

The first studies performed with the ReactorSTM concerned the prototype reaction of CO oxidation over Pt(110) [6, 61–63]. CO oxidation is one of the reactions taking place in the three-way automotive catalyst [64, 65]. Figure 1.10 shows characteristic STM images for CO oxidation over Pt(110) under reaction conditions [63]. The total pressure in the experiment was 1 bar, and the surface temperature 433 K. Figure 1.10a shows the initial (1 × 2) missing-row reconstruction, imaged in vacuum and at room temperature. This reconstruction is well known for Pt(110) under vacuum conditions [66]. When introducing 1 bar of CO in the reactor, and raising the temperature to 433 K, we find initially increased roughness (see Fig. 1.10b), which decays over time, resulting in the flat (1 × 1) surface (see Fig. 1.10c). After adding O_2 to the flow in a $CO:O_2$ ratio ≤ 0.2 (i.e. O_2-rich) the (1 × 1) surface changes to a

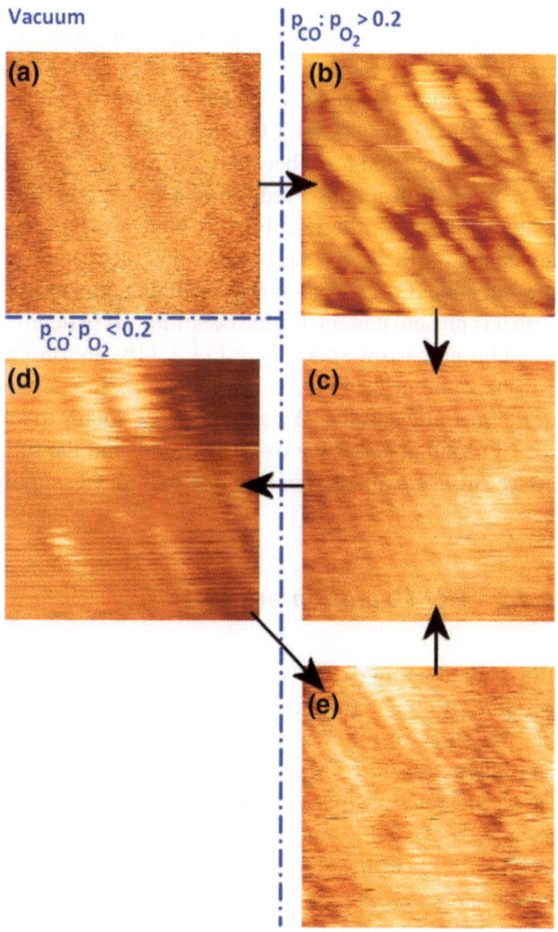

Fig. 1.10 STM images demonstrating the development of roughness at various stages of CO oxidation. **a** missing-row reconstruction, room temperature, vacuum, 4.5 nm × 4.5 nm, V_{bias} = −0.10 V, I_{tunnel} = 52 pA; **b** lifting of the missing-row reconstruction, T = 433 K, 1 bar CO, 15 nm × 15 nm, V_{bias} = 0.10 V, I_{tunnel} = 749 pA; **c** flat (1 × 1) structure in CO-rich flow, T = 433 K, 4.5 nm × 4.5 nm, V_{bias} = −0.04 V, I_{tunnel} = 86 pA; **d** commensurate (1 × 2) structure, observed immediately after switching to a more O_2-rich gas mixture, T = 433 K, 4.5 nm × 4.5 nm, V_{bias} = −0.04 V, I_{tunnel} = 86 pA; **e** Rough, metallic (1 × 1) surface, observed after increasing the CO content of the gas mixture again, T = 433 K, 4.5 nm × 4.5 nm, V_{bias} = 0.08 V, I_{tunnel} = 1004 pA. This figure was reprinted from [63] with permission from Elsevier

(1 × 2) structure. This is shown in Fig. 1.10d. No immediate increase in roughness is observed here, excluding that this structure reflects the formation of the (1 × 2) missing-row reconstruction. When a Pt(110) surface switches from the (1 × 1) structure to the (1 × 2) missing-row reconstruction, the number of Pt atoms present in the top layer is halved. As a result, after the transition, the surface is expected to exhibit an island-and-valley pattern of two height levels. This transition-induced roughness would then decay over time. Prolonged exposure of Pt(110) to an O_2-rich mixture results in an increase in roughness over time, which is related to catalytic turn-over. This will be discussed in more detail below. The result of this roughening is still present immediately after switching the gas composition to CO-rich conditions. Figure 1.10e shows the rough, metallic (1 × 1) surface, characteristic for a CO-rich gas mixture. The roughness of the surface decreases steadily over time to the level of Fig. 1.10c. The increasing roughness under reaction conditions has been observed before on Pt(110) [6], and was attributed to a Mars-van-Krevelen-like reaction mechanism [67]. An essential ingredient in this scenario is that the (1 × 2) structure observed under O_2-rich reaction conditions should be regarded as an ultrathin, epitaxial surface oxide. This has been inferred from the first STM observations [6] and it has been verified by Surface X-Ray Diffraction measurements under similar reaction conditions and accompanying Density Functional Theory calculations [8].

The kinetic results for CO oxidation on Pt(110) are shown in Fig. 1.11. When switching back and forth between CO-rich and O_2-rich mixtures, broad peaks in the CO_2 production are observed. These peaks are indicative of Langmuir-Hinshelwood kinetics, in which the reactivity is highest when the CO and O surface coverages are equal. Furthermore, we observe that the reactivity of the surface oxide, which is present under O_2-rich conditions, is higher than the reactivity of the metallic surface, which is present under CO-rich conditions (see details in panels b and c of Fig. 1.11). In addition, narrow spikes (see Fig. 1.11b) are observed just before the Langmuir-Hinshelwood peaks in the CO_2 production, but only when switching from the oxide to the metallic surface. We ascribe these spikes to an increased reaction rate when the CO content above the oxide is raised, and the sudden drop in reaction when the active surface oxide is removed due to high CO supply.

1.4.2 CO Oxidation on Pd(100)

CO oxidation on Pd(100) was investigated at a total pressure of 1.25 bar and a temperature of 408 K using the ReactorSTM [61, 62, 68]. Figure 1.12 shows the partial pressures of the reactants (CO and O_2) and the product (CO_2) (upper panel) and characteristic STM images (lower panel). At $t = 0$ s the CO pressure was slowly decreased. Image A shows the metallic Pd(100) surface, consisting of flat terraces separated by monoatomic steps. This surface has the same structure as a clean Pd(100) surface imaged in UHV. When decreasing the CO pressure, the CO_2 pressure decreases accordingly. However, at $t = 1612$ s, a sudden increase in CO_2 production is observed, coinciding with a significant change of the structure of the

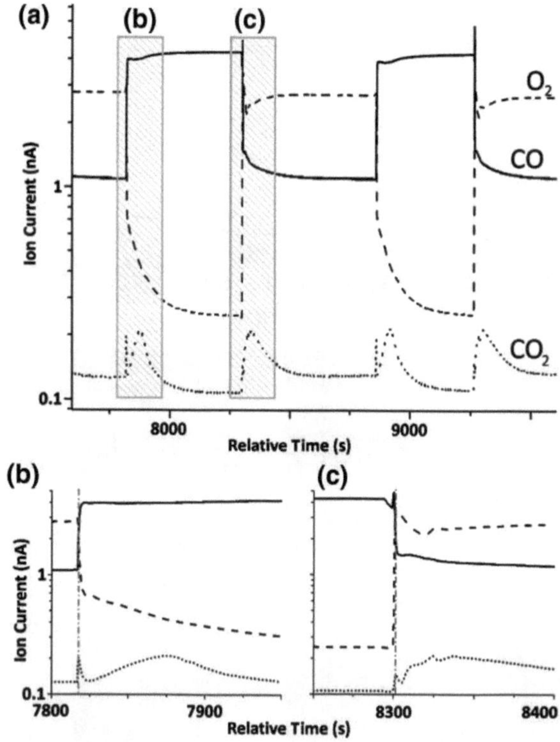

Fig. 1.11 Semi-logarithmic plot of the reaction kinetics during CO oxidation on Pt (110) at 1 bar and 433 K, measured with a QMS. **a** Two cycles in which the gas composition has been changed back and forth from CO-rich (*solid line*) to O_2-rich (*dashed line*). The product, CO_2, (*dotted line*) shows broad peaks during the switching that are indicative of Langmuir-Hinshelwood kinetics. **b** and **c** show zooms of the *gray dashed regions*. The oxide phase shows higher activity. The spike in **b** indicates the high reaction rate on the oxide during the initial stage of increase of the CO partial pressure, followed by the drop in the rate, when the oxide phase is removed due to the high CO partial pressure. This figure was reprinted from [63] with permission from Elsevier

Pd(100) surface (Image B, top). A large number of clusters is observed, with heights equal to that of a monoatomic step of Pd(100). This formation of clusters is found to accompany the initial stages of Pd(100) oxidation [69]. Therefore, we conclude that in Image B the surface oxidizes to form PdO, and that this causes the observed increase in reactivity. In time, roughness develops and a polycrystalline, granular structure is formed (Images C to E). In this stage, the CO_2 production decreases proportional to the decreasing CO pressure. At $t = 4700$ s we increase the CO partial pressure again. The Pd smoothens and the flat, metallic surface is restored (Images F and G). With the increase in CO pressure, a modest increase in CO_2 production is observed. At $t \sim 6700$ s we switch off the O_2 flow and we increase the CO pressure (Image H). Switching from an oxygen-rich to a CO-rich flow, the CO_2 production peaks at $t = 7010$ s. As the O_2 was pumped out, the CO_2 production steadily

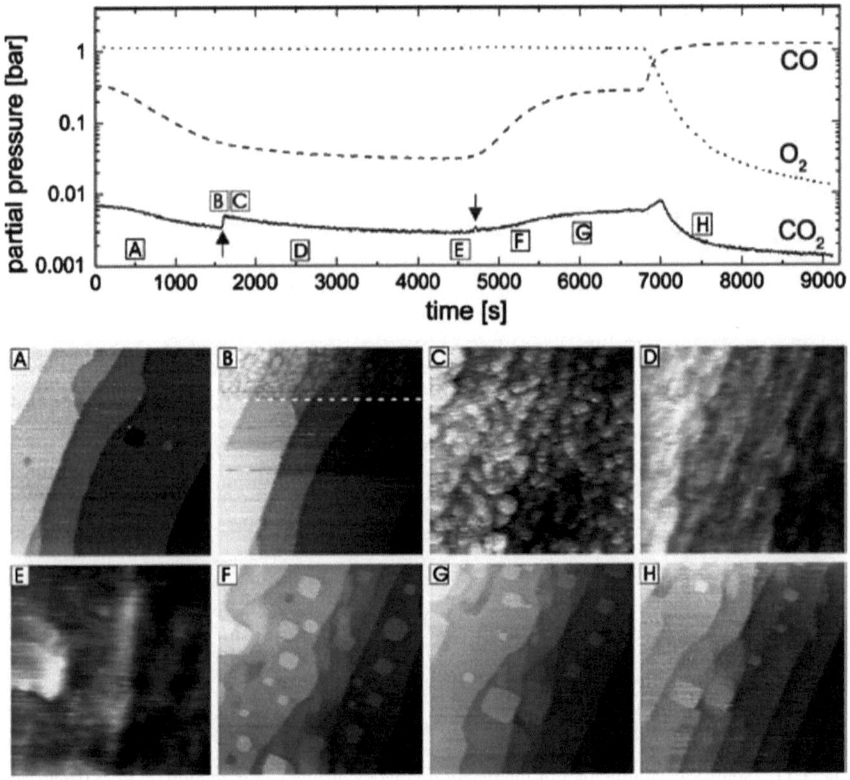

Fig. 1.12 (*Upper panel*) Mass spectrometer signals of the reactants (CO and O_2) and the product (CO_2) during CO oxidation on the Pd(100) surface at a total pressure of 1.25 bar and a temperature of 408 K. The labels A–H refer to the STM images shown in the *lower panel*. The *arrows* mark the changes in reactivity from low to high and vice versa. (*Lower panel*) STM images (140 nm x 140 nm, V_{bias} = 112 mV, I_{tunnel} = 200 pA) obtained during the CO oxidation reaction. This figure was reprinted from [68] with permission from Elsevier

decreased. When comparing Images G and H, we observe no essential structural changes when switching from a modest to a high ratio between the partial pressures of CO and O_2.

1.4.3 NO Reduction on Pt(110)

The catalytic conversion of nitrogen oxides (NO_x) is one of the three processes taking place on the three-way car catalyst [64, 65]. The major exhaust pollutants of cars are hydrocarbons, CO, and NO_x. The NO_x is formed during the phase of very high temperature (>1500 °C) of the combustion process resulting in thermal fixation of the nitrogen in air [70]. The exhaust of NO_x and hydrocarbons into the air, combined with sun light, results in the formation of ozone, a major component of smog. It was estimated that by the year 2000, over 800 million tons of combined pollutants of hydrocarbons, CO, and NO_x have been abated using automotive catalysts [71].

Considering the importance of the NO reduction reaction (either by H_2 or CO), it is very surprising that almost no *in situ* or *operando* studies have been performed for this system. To the best of our knowledge, the only results for NO reduction on platinum under reaction conditions have been obtained in our group [72]. In this work it was shown that the results for the NO + CO reaction on Pt(100) obtained under high-pressure conditions strongly differ from results obtained in UHV. In the former case, the Pt(100) surface switches between a (1×1) reconstruction and a quasi-hexagonal structure, depending on the CO/NO ratio at 1.25 bar [72]. Here we discuss the reduction of NO by H_2 over Pt(110). This reaction is not very selective over Pt, and many products, e.g. NH_3, H_2O, N_2, and N_2O, can be formed.

Figure 1.13 shows the results when exposing the Pt(110) surface to H_2 and NO. In Fig. 1.13a the gas composition is shown as a function of time. Gray bars show the times at which the STM images were measured. Prolonged exposure to pure H_2 at room temperature at a pressure of 1.2 bar resulted in a row structure with a (1×4) periodicity (see Fig. 1.13b). Also, some deeper missing-row features can be observed. Subsequently, this surface has been exposed to NO at increasing NO:H_2 ratios (Fig. 1.13c–e). The blue dot in the first image (b) indicates the switching point from pure H_2 to a mixture with a NO:H_2 ratio of 1:7.2. Exposure to NO slowly lifts the missing-row reconstruction, leaving flat terraces on the surface.

1.4.4 Fischer-Tropsch Synthesis on Co(0001)

In the Fischer-Tropsch synthesis reaction [73] a mixture of H_2 and CO (syngas) is converted via a surface polymerization reaction over a cobalt-based catalyst into linear hydrocarbons [74]. Industrial conditions for this reaction include pressures of 1 to several tens of bars, and temperatures between 423 and 573 K. Figure 1.14a shows an STM image obtained in a CO:H_2:Ar = 1:2:2 mixture at 4 bar and 483 K over Co(0001) [75]. From the image it is clear that under these reaction conditions the surface is completely covered by arrays of stripes, that run parallel to the $\langle 10\bar{1}0 \rangle$ crystallographic directions of the Co(0001) surface. The stripe pattern exhibits a period of 1.8 ± 0.3 nm, which is the same for the three equivalent orientations and is found everywhere on the surface. The period can be recognized easily in Fig. 1.14b, which is a zoomed-in part of Fig. 1.14a. The stripe pattern is strongly reminiscent of similar patterns reported in the literature for e.g. $C_{16}H_{34}$ on Au(111) [76] and $C_{14}H_{30}$ on highly-oriented pyrolithic graphite (HOPG) [77]. These hydrocarbon molecules self-assemble into regular, striped patterns. We therefore interpret the striped pattern observed in our experiments as hydrocarbon molecules synthesized in the Fischer-Tropsch reaction that lie flat on the Co(0001) surface. The period of the stripes L

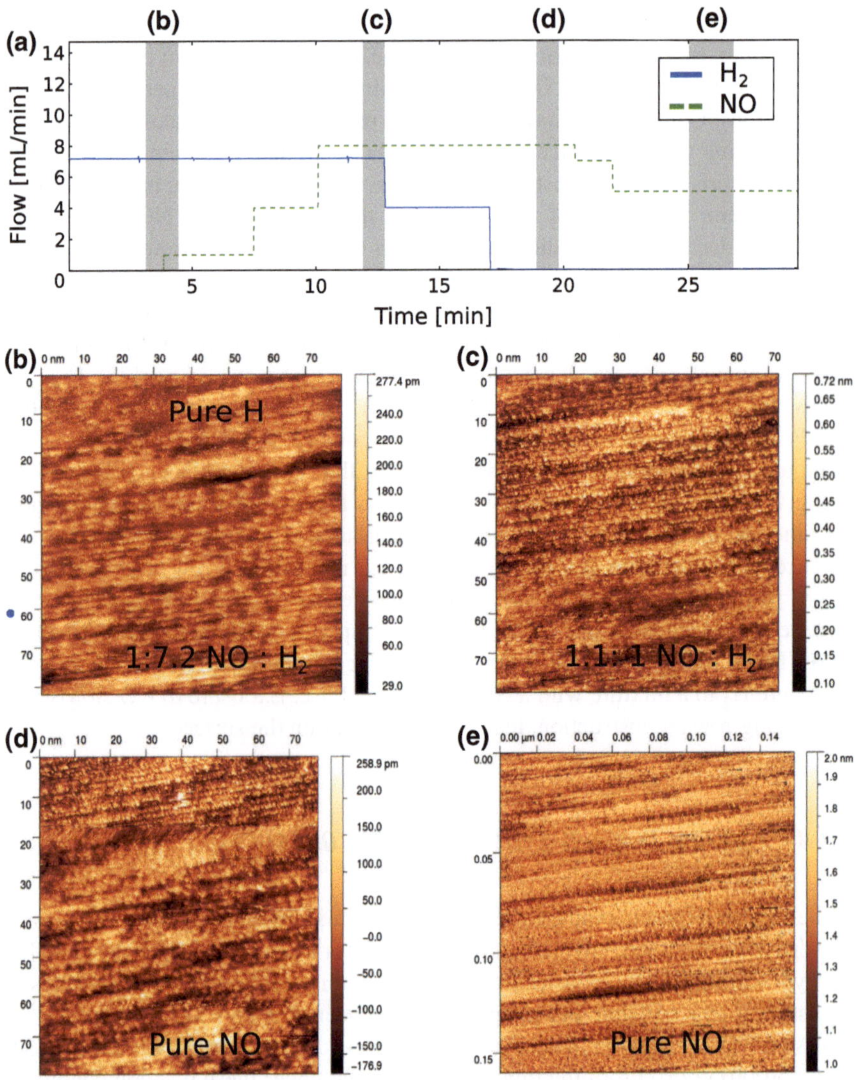

Fig. 1.13 Gas composition as a function of time and *in situ* STM images during NO reduction on Pt(110) at a total pressure of 1.2 bar and room temperature. **a** stepwise change from H_2 (*blue*) to NO (*green*) at room temperature and a total pressure of 1.2 bar. *Gray regions* indicate at which point in time under what conditions the STM images were obtained; **b** Pure H_2 followed by a switch to a partial pressure ratio of NO:H_2 = 1:7.2, 80 nm × 82 nm, $V_{bias} = -0.70$ V, $I_{tunnel} = 139$ pA; **c** Pt(110) at a partial pressure ratio NO:H_2 = 1.1 : 1.0, 73 nm × 80 nm, $V_{bias} = -0.70$ V, $I_{tunnel} = 141$ pA; **d** pure NO atmosphere, 78 nm × 80 nm, $V_{bias} = -0.70$ V, $I_{tunnel} = 141$ pA; **e** after prolonged NO exposure, 158 nm × 160 nm, $V_{bias} = -0.70$ V, $I_{tunnel} = 139$ pA. This figure was reprinted from [63] with permission from Elsevier

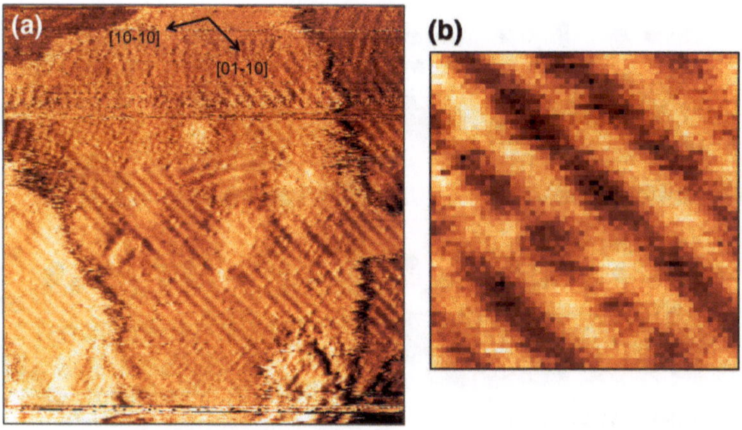

Fig. 1.14 STM image of **a** Co(0001) exposed to 4 bar of CO:H_2:Ar = 1:2:2 at 483 K, 62 nm × 62 nm; **b** enlarged view of part of (**a**), 7.5 nm × 7.5 nm. $V_{bias} = 0.3$ V, $I_{tunnel} = 100$ pA

is directly related to the molecular length l_n of a linear hydrocarbon with n carbon atoms [78, 79]: $L = l_n = l_0 + n \times l_{CH_2}$. Here, l_{CH_2} stands for the length of a CH_2 unit and l_0 is a length offset that takes into account the two methyl end groups of the hydrocarbon chains and the distance between the molecules that face each other in neighboring stripes. The observed stripe period corresponds to a molecular chain length of 14 ± 2 C atoms.

We attribute the emergence of a well-defined period to the combination of three processes that take place on the surface. The first process is the formation of hydrocarbons on the surface via a chain growth mechanism: H_2 and CO dissociate on the Co(0001) surface, and react to form hydrocarbons and water. The length of the hydrocarbons formed is a linear function of time: At each length, there is a constant probability for the chain growth process to continue, while the alternative is for the molecule to detach from the active site (step), at which point the chain stops growing. This scenario leads to an exponential distribution of produced chain lengths, which is dominated very strongly by the shorter chains. The second process is the thermally activated desorption of the hydrocarbons from the cobalt surface. When we assume the desorption energy to be proportional to the chain length, the short molecules stay on the surface for a very short time, while the longer ones tend to slowly accumulate. Finally, the third process is the condensation of the accumulating molecules into well-ordered, two-dimensional islands. This will happen when the surface is populated sufficiently strongly with molecules. The first molecular length that reaches the condensation level defines the stripe period. Shorter molecules do not reach a sufficiently high coverage, while longer molecules could in principle accumulate even more strongly, but would do so at such a late stage, that by then the surface is already saturated with a full molecular monolayer. Model calculations confirm this mechanism and show that a relatively long molecular chain length in the

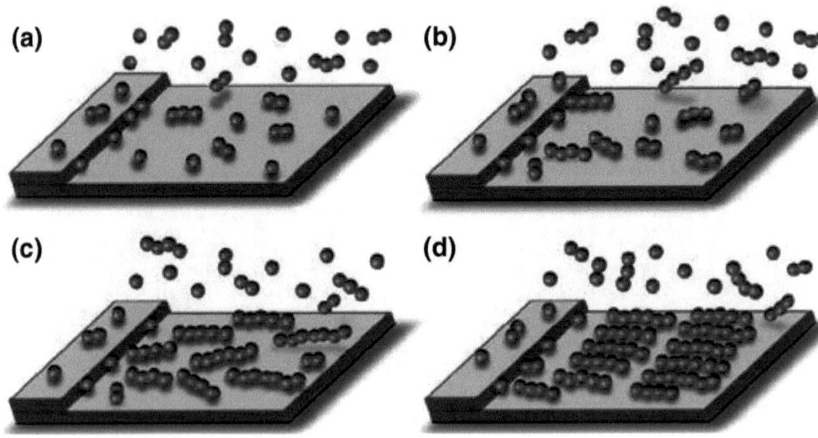

Fig. 1.15 Schematic scenario for the formation of the striped alkane phase on Co(0001) during the early stages of FT synthesis. The cobalt surface is shown as two terraces, separated by a monoatomic step. For the hydrocarbon chains, only the C atoms are depicted, as *green balls*. **a**, **b** Formation of monomers, dimers, and longer linear chains, growing from the step sites. **c**, **d** Desorption of the shorter chains from the surface and accumulation of longer chains on the cobalt terraces

order of 14 C atoms should indeed be expected. This process is qualitatively depicted in Fig. 1.15 [75].

1.4.5 Hydrodesulfurization Reaction

Hydrodesulfurization is the treatment of fossil fuels to remove sulfur-containing compounds from crude oil. Due to the stringent requirements for ultralow sulfur fuels, the interest in the catalytic hydrodesulfurization (HDS) process has grown. The industrial catalyst for HDS consists of alumina-supported MoS_2 nanoparticles. These nanoparticles are formed via sulfidation of MoO_3 nanoparticles. The active phase is generally accepted to be present as a few nm-wide, single-layer, MoS_2-like nanoparticles, usually promoted with Co or Ni [80]. The HDS activity has been attributed to sites at the edges of MoS_2 clusters, where S atoms are stripped off by H_2, creating sulfur vacancies. These vacancies are believed to have a high affinity for binding the sulfur-containing fuel molecules, thereby facilitating the sulfur extraction. Some of the basic steps in this process have been recognized in STM images obtained under UHV conditions (see e.g. [81–83]). Here we show the first STM results for HDS obtained under high-pressure conditions.

Figure 1.16a shows an STM image taken in UHV of MoS_2 nanoparticles grown on a Au(111) substrate, the model catalyst for HDS in our studies. As can be clearly seen, the MoS_2 particles form triangular shapes, truncated triangles, and hexagonal shapes, in agreement with previous literature, see e.g. [83–85]. In previous

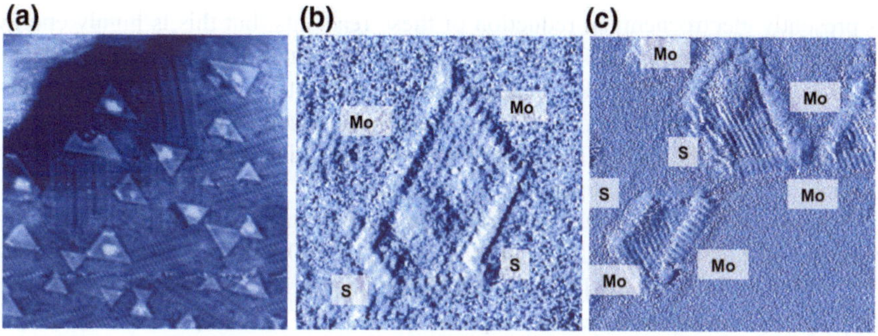

Fig. 1.16 **a** STM image taken in UHV of MoS_2 nanoparticles grown on a Au(111) substrate, 32 nm × 32 nm, $V_{bias} = -0.30$ V, $I_{tunnel} = 340$ pA; **b** STM image of MoS_2/Au(111) taken in 1 bar H_2 at 326 K, 7.5 nm × 7.5 nm, $V_{bias} = -0.30$ V, $I_{tunnel} = 675$ pA; **c** STM image of MoS_2/Au(111) taken under HDS reaction conditions: 1 bar 90% H_2, 10% CH_3SH at 523 K, 7.5 nm × 7.5 nm, $V_{bias} = -0.30$ V, $I_{tunnel} = 675$ pA. Images **b** and **c** are shown in derivative mode, to enhance contrast

UHV studies, it was found that the nanoparticles can have different structures at their edges [84]. Hexagonal nanoparticles will have edge terminations consisting of $(10\bar{1}0)$ Mo edges and $(\bar{1}010)$ S edges. The Mo edges are considered to be the most dominant sites and the active sites for the HDS reaction. Under reaction conditions, these Mo sites can have different coverages of S, which can be observed from the orientation of the edge atoms with respect to the atoms in the basal plane of the particle. When the Mo edge is fully covered with S, i.e. the 100% S-Mo edge, the atoms in the edge sites are out of registry with the atoms in the basal plane [84]. When a 50% S coverage is present, the Mo atoms in the edge are in registry with the Mo atoms in the basal plane of the particle.

When exposing the MoS_2 nanoparticles to 1 bar of H_2 at slightly elevated temperature (326 K), the nanoparticles form predominantly hexagonal shapes, as seen in Fig. 1.16b. In the image the presence of both Mo edges and S edges in one nanoparticle can be observed. From the registry of the edge Mo atoms with respect to the atoms in the basal plane, we conclude that we are observing a 50% S-Mo edge here. When exposing the MoS_2 nanoparticles to HDS conditions (90% H_2, 10% CH_3SH at 1 bar and 523 K), see Fig. 1.16c, we observe hexagonal nanoparticles. The particles have both S and Mo edges. The top two particles in Fig. 1.16c show 50% S-Mo edges, while the bottom particle shows 100% S-Mo edges.

1.4.6 Deacon Process for Chlorine Production

Chlorine is one of the most important compounds produced in the chemical industry. Its worldwide annual production is approximately 50 megaton, and it is responsible for approximately 50% of the total turnover of the chemical industry [86]. It is usually produced from HCl or chloride salts. The production method of choice is presently electrochemical reduction of these reactants, but this is highly energy demanding and it generates large quantities of CO_2. Due to these high economical costs and undesired effects for the environment, interest is growing in the heterogeneously catalyzed oxidation of HCl, the so-called Deacon process [87, 88]. Sumitomo Chemicals claims that the Deacon process can reduce the power consumption from 1100 kwh to 165 kwh per ton Cl_2 produced [89].

Several catalysts have been developed for the industrial Deacon process. Examples are $CuCl_2$ [87, 88], $CuCl_2$-KCl/SiO_2 [90, 91], Cr_2O_3/SiO_2 [92–94], RuO_2/TiO_2 [89, 95], and RuO_2/SnO_2 [96–98]. All these catalysts have been employed with varying degrees of success. Severe problems are loss of catalytic activity due to volatilization of the active phase, sintering of active sites and supports, and corrosion issues due to the combined presence of HCl and H_2O. At the moment, the catalysts employed in the industry (i.e. Sumitomo Chemicals and Covestro) are based on RuO_2. To obtain fundamental insight in the delicate interplay between the atomic-scale structure, the reaction mechanisms, and the performance of the catalyst, we are currently investigating its behavior under relevant conditions of high pressure and high temperature. Here we present some first results and we discuss the experimental issues of this highly aggressive system.

Figure 1.17a, b shows STM images of the $RuO_2(110)/Ru(0001)$ model catalyst obtained under UHV conditions. The images show rectangular-shaped terraces (a) and vaguely visible oxide rows (b), well-known for this surface [5, 99]. Upon exposure to the reaction mixture (1 bar of $HCl{:}O_2 = 1{:}4$) at room temperature (Fig. 1.17c), holes appear in the surface, due to etching by HCl. At this temperature, it seems that no re-oxidation of the film occurs when switching to 1 bar of O_2. Within the sensitivity limits of our QMS, no production of Cl_2 is observed at room temperature. X-ray photoelectron spectra taken after the reaction, however, show the presence of Cl on the surface.

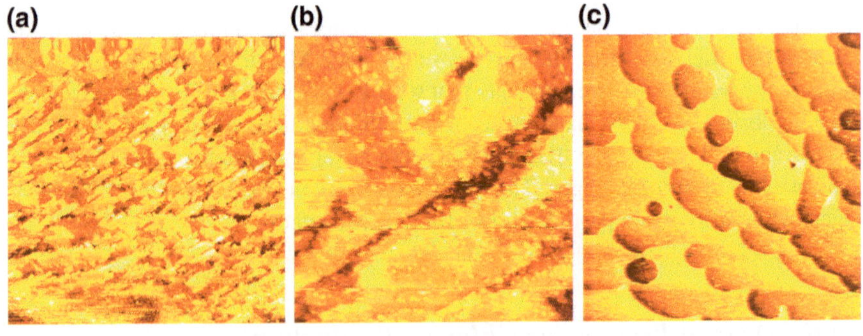

Fig. 1.17 **a** STM image of a $RuO_2(110)$ film on Ru(0001) under UHV conditions at room temperature, 400 nm × 400 nm, $V_{bias} = 3.0$ V, $I_{tunnel} = 300$ pA; **b** STM image of a $RuO_2(110)$ film on Ru(0001) under UHV conditions at room temperature, 50 nm × 50 nm, $V_{bias} = 3.0$ V, $I_{tunnel} = 250$ pA; **c** STM image of a $RuO_2(110)$ film on Ru(0001) during exposure to 1 bar of $HCl{:}O_2 = 1{:}4$ at room temperature, 800 nm × 800 nm, $V_{bias} = 2.75$ V, $I_{tunnel} = 100$ pA

With these first images we have shown that our set-up is capable of dealing with these very aggressive reaction conditions. However, some experimental issues were observed and have been dealt with. These include corrosion of the W tip (we now switched to PtIr), corrosion of the Cu seal in the leak valve of the QMS chamber (we now switched to a teflon-type seal), and decomposition of the Kalrez seal due to the combination of elevated temperatures and corrosive gases.

1.5 Conclusions and Outlook

Using the ReactorSTM and ReactorAFM we are able to cast a direct view on catalyst surfaces under industrially relevant conditions of high pressure and temperature. The ReactorSTM is now in a sufficiently mature state that reactions more complicated than CO oxidation can be studied. As is clear from the examples discussed here, we find pressure gaps for most reaction systems, simply by observing surface structures under reaction conditions that have been observed never before in UHV. The effect of the high pressure is to thermodynamically or kinetically stabilize new surface structures and compositions that incorporate one or more of the reactants. These observations show that a catalyst is more than a mere spectator in chemical reactions: It actively participates! The first results for aggressive chemical reactions, such as hydrodesulfurization and chlorine production, have given us confidence that the instruments described in this Chapter are able to perform under extremely corrosive conditions, showing no signs of degradation or destruction. Preliminary experiments using the ReactorAFM have shown that we can perform *in situ* studies of non-conducting surfaces under reaction conditions. With the promise of this instrument enabling us to obtain atomic-scale images of metallic nanoparticles supported on oxidic, insulating surfaces, we will be able in the near future to also bridge the materials gap.

Acknowledgements The work described in this chapter would not have been possible without the contributions of many people. The authors would like to thank Rik Mom, Violeta Navarro, Willem Onderwaater, Sander Roobol, Matthijs van Spronsen, Johan Bakker, Mirthe Bergman, Stefania Bobaru, Marta Cañas–Ventura, Bas Hendriksen, Kees Herbschleb, Qian Liu, Peter Rasmussen, Marcel Rost, and Dunja Stolz. The authors are indebted to the fine-mechanical department of Leiden University, and especially to Peter van der Tuijn, who was vital for the design of the latest version of the ReactorSTM and the ReactorAFM. We owe a special thanks to Gertjan van Baarle and Alexei Ofitserov of Leiden Probe Microscopy B.V., who participated in the design of the prototypes of the Reactor set-ups and developed the commercial version of the ReactorSTM. This work is financially supported by NanoNextNL, a micro- and nanotechnology consortium of the Government of the Netherlands and 130 partners, and by a SmartMix grant and the NIMIC partner organizations through NIMIC, a public-private program. IMNG acknowledges the support of STW, which is financially supported by the Netherlands Organization for Scientific Research (NWO), through a Veni grant.

References

1. J. Wintterlin, Adv. Catal. **45**, 131 (2000)
2. P. Stoltze, J.K. Nørskov, Phys. Rev. Lett. **55**, 2502 (1985)
3. J.A. Rodriguez, D.W. Goodman, Surf. Sci. Rep. **14**, 1 (1991)
4. X. Su, P.S. Cremer, Y.R. Shen, G.A. Somorjai, J. Am. Chem. Soc. **119**, 3994 (1997)
5. H. Over, Y.D. Kim, A.P. Seitsonen, S. Wendt, E. Lundgren, M. Schmid, P. Varga, A. Morgante, G. Ertl, Science **287**, 1474 (2000)
6. B.L.M. Hendriksen, J.W.M. Frenken, Phys. Rev. Lett. **89**, 046101 (2002)
7. H. Over, M. Muhler, Prog. Surf. Sci. **72**, 3 (2003)
8. M.D. Ackermann, T.M. Pedersen, B.L.M. Hendriksen, O. Robach, S.C. Bobaru, I. Popa, C. Quiros, H. Kim, B. Hammer, S. Ferrer, J.W.M. Frenken, Phys. Rev. Lett. **95**, 25505 (2005)
9. R. Westerström, J. Gustafson, A. Resta, A. Mikkelsen, J.N. Andersen, E. Lundgren, N. Seriani, F. Mittendorfer, M. Schmid, J. Klikovits, P. Varga, M.D. Ackermann, J.W.M. Frenken, N. Kasper, A. Stierle, Phys. Rev. B **76**, 155410 (2007)
10. H. Bluhm, M. Hävecker, A. Knop-Gericke, M. Kiskinova, R. Schlögl, M. Salmeron, MRS Bull. **32**, 1022 (2007)
11. M. Salmeron, R. Schlögl, Surf. Sci. Rep. **63**, 169 (2008)
12. W.P.A. Jansen, A.W.D. van der Gon, G.M. Wijers, Y.G.M. Rikers, H.H. Brongersma, P.W. van der Hoogen, J.A.M. de Laat, T.M. Maas, E.C.A. Dekkers, P. Brinkgreve, Rev. Sci. Instrum. **73**, 354 (2002)
13. P.L. Hansen, J.B. Wagner, S. Helveg, J.R. Rostrup-Nielsen, B.S. Clausen, H. Topsøe, Science **295**, 2053 (2002)
14. P.L. Gai, E.D. Boyes, S. Helveg, P.L. Hansen, S. Giorgio, C. Henry, MRS Bull. **32**, 1044 (2007)
15. J.F. Creemer, S. Helveg, G.H. Hoveling, S. Ullmann, A.M. Molenbroek, P.M. Sarro, H.W. Zandbergen, Ultramicroscopy **108**, 993 (2008)
16. E. de Smit, I. Swart, J.F. Creemer, G.H. Hoveling, M.K. Gilles, T. Tyliszczak, P.J. Kooyman, H.W. Zandbergen, C. Morin, B.M. Weckhuysen, F.M.F. de Groot, Nature **456**, 222 (2008)
17. P.B. Rasmussen, B.L.M. Hendriksen, H. Zeijlemaker, H.G. Ficke, J.W.M. Frenken, Rev. Sci. Instrum. **69**, 3879 (1998)
18. B.L. Weeks, C. Durkan, H. Kuramochi, M.E. Welland, T. Rayment, Rev. Sci. Instrum. **71**, 3777 (2000)
19. F. Tao, D. Tang, M. Salmeron, G.A. Somorjai, Rev. Sci. Instrum. **79**, 084101 (2008)
20. C.T. Herbschleb, P.C. van der Tuijn, S.B. Roobol, V. Navarro, J.W. Bakker, Q. Liu, D. Stoltz, M.E. Cañas-Ventura, G. Verdoes, M.A. van Spronsen, M. Bergman, L. Crama, I. Taminiau, A. Ofitserov, G.J.C. van Baarle, J.W.M. Frenken, Rev. Sci. Instrum. **85**, 083703 (2014)
21. S.B. Roobol, M.E. Cañas-Ventura, M. Bergman, M.A. van Spronsen, W.G. Onderwaater, P.C. van der Tuijn, R. Koehler, A. Ofitserov, G.J.C. van Baarle, J.W.M. Frenken, Rev. Sci. Instrum. **86**, 033706 (2015)
22. X. Su, P.S. Cremer, Y.R. Shen, G.A. Somorjai, Phys. Rev. Lett. **77**, 3858 (1996)
23. G. Rupprechter, T. Dellwig, H. Unterhalt, H.-J. Freund, J. Phys. Chem. B **105**, 3797 (2001)
24. G. Rupprechter, MRS Bull. **32**, 1031 (2007)
25. G.A. Beitel, A. Laskov, H. Oosterbeek, E.W. Kuipers, J. Phys. Chem **100**, 12494 (1996)
26. C. Li, P.C. Stair, Catal. Lett. **36**, 119 (1996)
27. P.C. Stair, C. Li, J. Vac. Sci. Technol. A **15**, 1679 (1997)
28. C. Li, P.C. Stair, Catal. Today **33**, 353 (1997)
29. H.H. Rotermund, Surf. Sci. **386**, 10 (1997)
30. H. Isern, K. Peters, P. Steadman, O. Robach, J. Alvarez, E. Lundgren, S. Ferrer, Surf. Sci **482–485**, 101 (2001)
31. S. Ferrer, M.D. Ackermann, E. Lundgren, MRS Bull. **32**, 1010 (2007)
32. R. van Rijn, M.D. Ackermann, O. Balmes, T. Dufrane, A. Geluk, H. Gonzalez, H. Isern, E. de Kuyper, L. Petit, V.A. Sole, D. Wermeile, R. Felici, J.W.M. Frenken, Rev. Sci. Instrum. **81**, 014101 (2010)

33. A. Knop-Gericke, M. Hävecker, T. Schedel-Niedrig, R. Schlögl, Catal. Lett. **66**, 215 (2000)
34. J. Evans, A. Puig-Molina, M. Tromp, MRS Bull. **32**, 1038 (2007)
35. C.J. Chen, *Introduction to Scanning Tunneling Microscopy* (Oxford University Press, Oxford, 1993)
36. L. Kuipers, R.W.M. Loos, H. Neerings, J. ter Horst, G.J. Ruwiel, A.P. de Jongh, J.W.M. Frenken, Rev. Sci. Instrum. **66**, 4557 (1995)
37. M.S. Hoogeman, D.G. van Loon, R.W.M. Loos, H.G. Ficke, E. de Haas, J.J. van der Linden, H. Zeijlemaker, L. Kuipers, M.F. Chang, M.A.J. Klik, J.W.M. Frenken, Rev. Sci. Instrum. **69**, 2072 (1998)
38. G. Binnig, H. Rohrer, Helv. Phys. Acta. **55**, 726 (1982)
39. G. Binnig, H. Rohrer, Rev. Mod. Phys. **56**, 615 (1987)
40. G. Binnig, H. Rohrer, C. Gerber, E. Weibel, Appl. Phys. Lett. **40**, 178 (1982)
41. G. Binnig, H. Rohrer, C. Gerber, E. Weibel, Phys. Rev. Lett. **49**, 57 (1982)
42. R. Wiesendanger, *Scanning Probe Microscopy and Spectroscopy* (Cambridge University Press, Cambridge, 1994)
43. B.J. McIntyre, M. Salmeron, G.A. Somorjai, Rev. Sci. Instrum. **64**, 687 (1993)
44. J.A. Jensen, K.B. Rider, Y. Chen, M. Salmeron, G.A. Somorjai, J. Vac. Sci. Technol. B **17**, 1080 (1999)
45. E. Laegsgaard, L. Österlund, P. Thostrup, P.B. Rasmussen, I. Stensgaard, F. Besenbacher, Rev. Sci. Instrum. **72**, 3537 (2001)
46. A. Kolmakov, D.W. Goodman, Rev. Sci. Instrum. **74**, 2444 (2003)
47. M. Rössler, P. Geng, J. Wintterlin, Rev. Sci. Instrum. **76**, 023705 (2005)
48. F. Tao, L. Nguyen, S. Zhang, Rev. Sci. Instrum. **84**, 034101 (2013)
49. J. Lievonen, K. Ranttila, M. Ahlskog, Rev. Sci. Instrum. **78**, 043703 (2007)
50. D. D'Agostino, D. Jay, H. McNally, Microsc. Microanal. **16**, 1042 (2010)
51. S.R. Higgins, C.M. Eggleston, K.G. Knauss, C.O. Boro, Rev. Sci. Instrum. **69**, 2994 (1998)
52. A.S. Lea, S.R. Higgins, K.G. Knauss, K.M. Rosso, Rev. Sci. Instrum. **82**, 043709 (2011)
53. G. Dong, E.B. Fourré, F.C. Tabak, J.W.M. Frenken, Phys. Rev. Lett. **104**, 096102 (2010)
54. M. Rost, L. Crama, P. Schakel, E. van Tol, G. van Velzen-Williams, C. Overgauw, H. ter Horst, H. Dekker, B. Okhuijsen, M. Seynen, A. Vijftigschild, P. Han, A.J. Katan, K. Schoots, R. Schumm, W. van Loo, T.H. Oosterkamp, J.W.M. Frenken. Rev. Sci. Instrum. **76**, 053710 (2005)
55. F.J. Giessibl, S. Hembacher, H. Bielefeldt, Science **289**, 422 (2000)
56. L. Gross, F. Mohn, N. Moll, P. Liljeroth, G. Meyer, Science **325**, 1110 (2009)
57. F.J. Giessibl, Appl. Phys. Lett. **73**, 3956 (1998)
58. A. Botman, M. Hesselberth, J.J.L. Mulders, Microelectron. Eng. **85**, 1139 (2008)
59. R. Grober, J. Acimovic, J. Schuck, D. Hessman, P.J. Kindlemann, J. Hespanha, A.S. Morse, Rev. Sci. Instrum. **71**, 2776 (2000)
60. Leiden Probe Microscopy B.V. http://www.leidenprobemicroscopy.com/
61. B.L.M. Hendriksen, S.C. Bobaru, J.W.M. Frenken, Catal. Today **105**, 234 (2005)
62. B.L.M. Hendriksen, S.C. Bobaru, J.W.M. Frenken, Top. Catal. **36**, 43 (2005)
63. M.A. van Spronsen, G.J.C. van Baarle, C.T. Herbschleb, J.W.M. Frenken, I.M.N. Groot, Catal. Today **244**, 85 (2015)
64. R.M. Heck, R.J. Farrauto, Appl. Catal. A: General **221**, 443 (2001)
65. M. Shelef, Catal. Rev. Sci. Eng. **11**, 1 (1975)
66. E. Vlieg, I.K. Robinson, K. Kern, Surf. Sci. **233**, 248 (1990)
67. P. Mars, D.W. van Krevelen, Chem. Eng. Sci. **3**, 41 (1954)
68. B.L.M. Hendriksen, S.C. Bobaru, J.W.M. Frenken, Surf. Sci. **552**, 229 (2004)
69. G. Zheng, E.I. Altman, Surf. Sci. **504**, 253 (2002)
70. J. Zeldovich, Acta Phys. USSR **21**, 577 (1946)
71. R.M. Heck, R.J. Farrauto, Automotive Engineering February p. 93 (1996)
72. C.T. Herbschleb, S.C. Bobaru, J.W.M. Frenken, Catal. Today **154**, 61 (2010)
73. F. Fischer, H. Tropsch, Brennst. Chem. **7**, 97 (1926)
74. D. Leckel, Energy Fuels **23**, 2342 (2009)

75. V. Navarro, M.A. van Spronsen, J.W.M. Frenken, Nat. Chem. **8**, 929 (2016)
76. J. Cousty, A. Marchenko, Surf. Sci. **520**, 128 (2002)
77. Q. Chen, H.-J. Yan, C.-J. Yan, G.-B. Pan, L.-J. Wan, G.-Y. Wen, D.-Q. Zhang, Surf. Sci. **602**, 1256 (2008)
78. J. Marchenko, A. Cousty, Surf. Sci. **520**, 128 (2002)
79. K. Uosaki, R. Yamada, J. Am. Chem. Soc. **121**, 4090 (1999)
80. H. Topsøe, B.S. Clausen, F.E. Massoth, *Hydrotreating Catalysis, Science and Technology* (Springer Verlag, Berlin, 1996)
81. S. Helveg, J.V. Lauritsen, E. Laegsgaard, I. Stensgaard, J.K. Nørskov, B.S. Clausen, H. Topsøe, F. Besenbacher, Phys. Rev. Lett. **84**, 951 (2000)
82. J.V. Lauritsen, S. Helveg, E. Laegsgaard, I. Stensgaard, B.S. Clausen, H. Topsøe, F. Besenbacher, J. Catal. **197**, 1 (2001)
83. J.V. Lauritsen, M.V. Bollinger, E. Laegsgaard, K.W. Jacobsen, J.K. Nørskov, B.S. Clausen, H. Topsøe, F. Besenbacher, J. Catal. **221**, 510 (2004)
84. A.S. Walton, J.V. Lauritsen, H. Topsøe, F. Besenbacher, J. Catal. **308**, 306 (2013)
85. A. Tuxen, J. Kibsgaard, H. Gobel, E. Laegsgaard, H. Topsøe, J.V. Lauritsen, F. Besenbacher, ACS Nano **4**, 4677 (2010)
86. Eurochlor, Chlorine Industry Review (2007–2008)
87. H. Deacon, U.S. Patent **85370** (1868)
88. H. Deacon, U.S. Patent **165802** (1875)
89. K. Iwanaga, K. Seki, T. Hibi, K. Issoh, T. Suzuta, M. Nakada, Y. Mori, T. Abe, Sumitomo Kagaku **2004-I** (2004)
90. A.J. Johnson, A.J. Cherniavsky, U.S. Patent **2542961** (1951)
91. S.O. Company, U.S. Patent **3210158** (1965)
92. T. Kiyoura, N. Fujimoto, M. Ajioka, T. Suzuki, Y. Kogure, K. Kanaya, T. Nagayama, European Patent **EP184413A** (1984)
93. H. Itoh, Y. Kono, M. Ajioka, S. Takenaka, M. Kataita, U.S. Patent **4803065** (1989)
94. T. Kiyoura, Y. Kogure, T. Nagayama, K. Kanaya, U.S. Patent **4822589** (1989)
95. T. Hibi, H. Nishida, H. Abekawa, U.S. Patent **5871707** (1999)
96. A. Wolf, L. Mleczko, O. F. Schlüter, S. Schubert, European Patent **EP2026905** (2006)
97. A. Wolf, J. Kintrup, O.F. Schlüter, L. Mleczko, European Patent **EP2027062** (2006)
98. A. Wolf, L. Mleczko, S. Schubert, O.F. Schlüter, European Patent **EP2027063** (2006)
99. M. Rössler, S. Günther, J. Wintterlin, J. Phys. Chem. C **111**, 2242 (2007)

Chapter 2
Ambient-Pressure X-ray Photoelectron Spectroscopy (APXPS)

Osman Karslıoğlu and Hendrik Bluhm

Abstract X-ray photoelectron spectroscopy (XPS) is a powerful technique for studying surfaces, including those of heterogeneous catalysts, through its ability to quantitatively analyze the elemental and chemical composition with high surface sensitivity. The understanding of heterogeneous catalytic processes under realistic conditions requires measurements at elevated pressures, far from the high-vacuum conditions under which the majority of XPS measurements are conducted. The investigation of surfaces using XPS at or near relevant pressures poses challenges due to scattering of electrons by gas molecules, which have been overcome through the development of ambient-pressure XPS (APXPS). In this chapter, we will review technical approaches for conducting XPS at higher pressures and discuss other experimental challenges that need to be addressed in APXPS investigations. At the end of the chapter, examples of APXPS experiments of CO oxidation over Ru and Pd, as well as the oxidation of other small hydrocarbons are shown.

2.1 Technique

2.1.1 Basics of XPS

XPS is a surface-sensitive chemical-analysis technique, which has become a common tool in laboratories all around the world. Its applications span a wide range of fields from catalysis to electronics. The technique relies on the distinct binding energies of core electrons in different elements. Qualitatively, binding energy is a measure of how tightly an electron is bound to an atom. From a fundamental perspective, one can start defining it by considering the photoemission process. For an isolated atom, energy conservation during photoemission can be expressed as:

O. Karslıoğlu · H. Bluhm (✉)
Lawrence Berkeley National Laboratory, Chemical Sciences Division,
1 Cyclotron Road, Berkeley, CA 94720, USA
e-mail: hbluhm@lbl.gov

J. Frenken and I. Groot (eds.), Operando *Research in Heterogeneous Catalysis*,
Springer Series in Chemical Physics 114, DOI 10.1007/978-3-319-44439-0_2

$$E_i^N + h\nu = E_f^{N-1} + E_{kin} \quad (2.1)$$

where $h\nu$ is the photon energy, E_{kin} is the kinetic energy of the photoelectron, and E_i and E_f are the initial (N particles) and final (N − 1 particles) energies of the system. After rearranging the terms and defining:

$$E_b \equiv E_f^{N-1} - E_i^N \quad (2.2)$$

the equation becomes:

$$E_b = h\nu - E_{kin} \quad (2.3)$$

where E_b is the binding energy of an electron in the atom. Equation 2.3 implies that the binding energy is referenced to vacuum (free electron at rest). For solids (condensed phases in general) it is customary to define the binding energy relative to the Fermi level, which serves as a more practical reference. For solids, 2.3 is modified to:

$$E_b = h\nu - E_{kin} - \phi \quad (2.4)$$

where ϕ is the work function of the surface wherever E_{kin} is measured at, in the case of XPS that of the electron analyzer. The equation holds only if the Fermi levels of the sample and analyzer are aligned, which is the default configuration.

The binding energy of an electron depends mainly on the type of atom that it is bound to and its quantum state (i.e. energy level, 1s, 2s, 2p, etc.). In addition, differences in the chemical environment of an element (i.e. its oxidation state) can also cause changes in binding energy. These changes are referred to as chemical shifts, and it is the power to resolve these subtle differences that makes XPS an excellent method for the investigation of surface chemistry.

In an XPS experiment, monochromatic X-rays irradiate the sample and excite electrons in a sample to unbound states (i.e. to vacuum) and the number of electrons is counted as a function of kinetic energy. The kinetic-energy axis is usually converted to binding energy using 2.4. Most X-ray photoelectron spectrometers are of the concentric hemispherical-analyzer type, where the photoelectrons that leave the sample are collected by an electrostatic lens system, which focuses the electrons onto the entrance aperture of the concentric hemispherical analyzer. The heart of the electron-energy analyzer consists of two concentric hemispheres, which are biased such that only electrons with a certain kinetic energy (i.e. "pass energy") reach the electron detector (e.g. channeltrons or a phosphor screen with camera), which is situated at the exit of the hemispherical analyzer.

2.1.2 Operating XPS at Elevated Pressure

There are two basic problems that need to be overcome in XPS experiments under non-vacuum conditions. The fundamental problem is the scattering of photoelectrons by gas molecules, which limits the distance that electrons can travel through the gas atmosphere. For instance, the inelastic mean free path (IMFP) of electrons in water vapor at 1 Torr for 100 eV kinetic-energy electrons is about 1 mm, much shorter than the path length of electrons in a spectrometer, which is typically on the order of a meter or more. The other issue is of technical nature and concerns discharging through the gas where high voltage differences are present, as in the electrostatic lens system or the electron detectors. The voltages applied to the lens elements in general increase with the kinetic energy of the electrons analyzed and at some point a critical electrical-field gradient between adjacent lens elements is reached which leads to a gas discharge. This problem can be addressed through the appropriate design of the electrostatic lens (e.g. large enough gaps between the lens elements), the use of lens tables that avoid large potential differences between adjacent lens elements, as well as efficient differential pumping. In the following we will address some of the design criteria for ambient-pressure photoelectron spectrometers.

2.1.2.1 Electron Scattering in Matter

Electrons are scattered very strongly in matter due to the nature of electron-electron interactions [1]. For electrons with kinetic energies (E_{kin}) below the ionization energy of the gas molecules (typically below 10 eV), the nature of the interaction is elastic, which means electrons change direction but do not lose energy. The effect of elastic scattering to signal attenuation depends on the experimental configuration. If the area of the sample illuminated by the X-rays is large compared to the area seen by the spectrometer, the effect will be small—electrons scattered out of the acceptance cone will be compensated by the electrons scattered into it. If the X-ray spot is small, this compensation will not happen and more signal will be lost. In any case, the effect of elastic scattering will be small in a typical APXPS experiment since scattering is predominantly in the forward direction, and most APXPS experiments detect photoelectrons with higher kinetic energies, where elastic scattering is less relevant.

For kinetic energies above the ionization energy threshold, electron-molecule collisions may result in energy loss for the electron, i.e. inelastic scattering. This is the principal signal-attenuation mechanism in APXPS. Electrons with kinetic energies around 50–100 eV are scattered most strongly and thus have the smallest IMFP. The electron scattering cross section decreases as E_{kin} increases beyond 100 eV. For a more detailed discussion of different scattering mechanisms, the reader may refer to [2].

Signal attenuation due to electron scattering in the gas becomes a problem when the signal-to-noise ratio becomes too small for the acquisition of a spectrum to be completed within a "reasonable" period of time. For electrons with a given kinetic energy the acquisition time depends not only on the gas pressure and the electrons' path length in the gas but also on other factors. The signal is proportional to the photon flux and the photoionization cross section of the investigated core level, whereas the signal-to-noise ratio decreases with the peak-width of the core level and the electron scattering cross section of the gas molecule. It also depends on the acquisition parameters of the analyzer such as the pass energy.

Assuming all other parameters (e.g. photon flux, analyzer transmission, etc.) are kept constant, one can estimate a practical maximum pressure based on a typical distance that electrons can travel in a gas and the distance between the sample and the differentially-pumped entrance aperture to the spectrometer lens. The inelastic mean free path in condensed matter for electrons with 100 eV kinetic energy is about a nanometer [3]. Gases are about 1000 times less dense than their condensed forms at atmospheric conditions. This means electrons with ~100 eV kinetic energy can travel around 1 μm in a gas at 1 atm, which would limit the entrance aperture size to the same order of magnitude to avoid pressure inhomogeneities at the sample surface (see below) and also requires a tightly-focused X-ray beam that impinges on the sample surface under a very shallow angle to avoid shadowing by the aperture. More practical aperture-sample distances are in the 100 μm range, but this leads to a pressure limit of ~2 orders of magnitude lower than 1 atm. This means that most APXPS experiments currently are carried out at pressures below 10 Torr and at sample—aperture distances of 0.1 to 1 mm.

The electrostatic lens needs to be at a low enough pressure for the electrons to travel without significant scattering, and the electron detector needs to be at pressures well below 10^{-6} Torr. Transitioning from a high-pressure region to a low-pressure region can be done in two ways. One is to introduce a gas-impermeable but electron-permeable membrane between the sample and the spectrometer, such as graphene. This provides an abrupt drop in pressure, from Torr range to ultrahigh vacuum (UHV), and would be an ideal solution in terms of reducing the size and cost of the spectrometer (see Fig. 2.1b). However, this approach has so far shown to be difficult to implement technically due to the stringent requirements on the mechanical stability of the ultrathin membranes. Nevertheless, there are ongoing efforts in this direction [4–7], which recently resulted in successful demonstrations at 1 bar pressure [8, 9]. In the following we will focus on the most-commonly used scheme to transition from the ambient pressures in the sample compartment to high vacuum in the electron detector, namely the use of differential pumping [10].

2.1.2.2 Separation of Sample Cell and Electron Spectrometer by Differential Pumping

The conventional solution to reducing the path length of electrons through the gas phase in the sample cell has been to decrease the pressure using differential-pumping

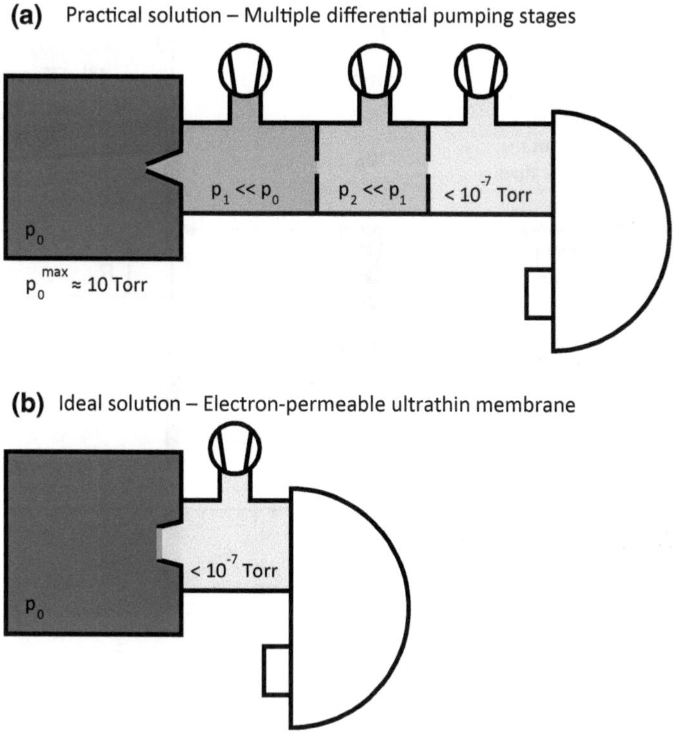

Fig. 2.1 Two possible schemes for an APXPS setup to transition from a high pressure at the sample to UHV at the electron detector. **a** Decreasing the pressure gradually using multiple differential pumping stages, **b** decreasing the pressure abruptly using an ultrathin membrane, which is permeable to electrons but impermeable to molecules

stages along the electron path. The differential-pumping stages are usually separated by apertures of the size on the order of a few millimeter, with the first aperture (facing the sample) of sub-millimeter dimensions. The apertures are small enough to limit gas flow but large enough to allow electron transmission. Figure 2.1a shows schematically how such a differential-pumping system works.

The differential-pumping system creates a pressure gradient along the electron trajectory. Molecular-flow equations for an aperture with negligible length indicate that most of the pressure drop occurs within two diameter-lengths around the aperture on the spectrometer axis. Figure 2.2a shows a sample placed at a distance "*z*" from the aperture, which has a radius of *R*. The calculated pressure map is plotted in part (b) of the figure. The details of the calculation can be found in [11]. Distances are given in units of *R*. The expression for pressure on the z-axis is:

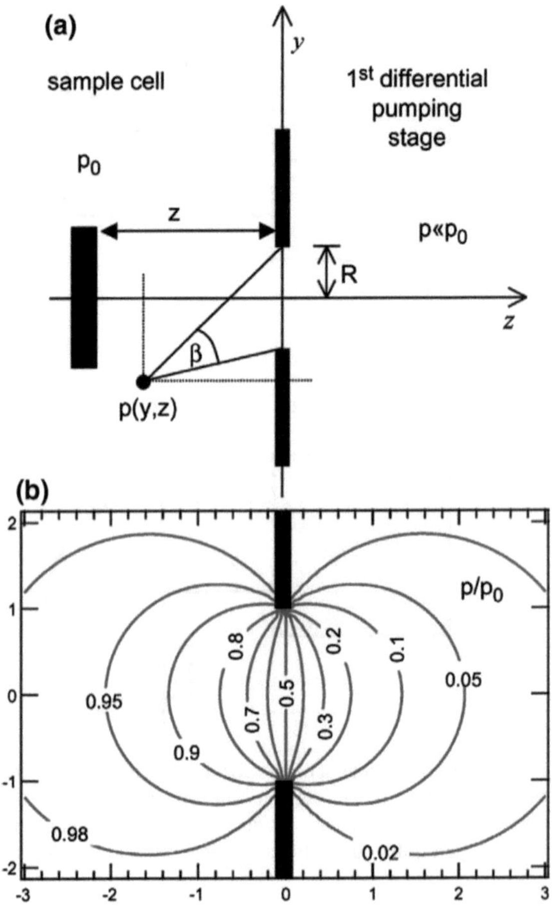

Fig. 2.2 Pressure near an aperture, which is differentially pumped on the right side. The pressure on the differentially-pumped side, sufficiently far away from the aperture, is much smaller than p_O, the pressure on the high-pressure side far away from the aperture. **a** Sample placed at a distance z from the plane of an aperture with radius R. **b** Isobars in the y-z plane. Pressures are in units of p_O and coordinates are in units of R. Reproduced with permission from [12]

$$p(z) = \frac{1}{2} p_0 \left(1 - \frac{z}{\sqrt{1+z^2}}\right) \tag{2.5}$$

This means that the electrons experience different pressures at different points along the path to the analyzer. For a simplified view one can assume that the electrons experience a pressure of p_O for a shorter path length, which we define as "effective length", and experience no gas pressure for the rest of the path. Effective length L for electrons emerging from point ($y = 0$, $z = z$) would then be defined as:

$$L(z) \equiv \frac{1}{p_0} \int_z^{\infty} p\left(z'\right) dz' = \frac{1}{p_0} \int_z^{\infty} \frac{1}{2} p_0 \left(1 - \frac{z'}{\sqrt{1+z'^2}}\right) dz' = \frac{1}{2} \int_z^{\infty} \left(1 - \frac{z'}{\sqrt{1+z'^2}}\right) dz' \tag{2.6}$$

For $z = 0$, $2R$, $4R$ and $6R$, L is $0.5R$, $2.1R$, $4.05R$, and $6.03R$, respectively. That means that for molecular-flow conditions the effective length is approximately the distance between the aperture and the sample.

Calculations for the case of a Knudsen-flow regime, where the mean free path of the molecules is comparable to the aperture size, require numerical simulations (an example can be found in [13]). Analytical expressions exist for the viscous-flow regime where the electron mean free path is much smaller than the aperture size [14]. In this case, the results are molecule specific. For water, the calculations yield an effective length of $1.03R$ after the aperture [11] (i.e. $L(0) = 1.03R$), as compared to $0.5R$ in the case of molecular flow. This is expected since the flow is more directional and pressure is higher on the analyzer axis, along the trajectories of the electrons. In practical terms, the difference of $L(0)$ of $0.53R$ between molecular and viscous-flow regimes is relatively small, i.e. in both cases the effective length is close to the distance of the sample from the aperture. However, for viscous flow the pressure-field gradients near the sample will be complex and dependent on the shape and size of the sample. Recently the pressure field between a planar sample and a differential-pumping aperture was studied using computational fluid dynamics (CFD) simulations in the turbulent-flow regime [15]. The results show that as the pressure in the chamber is increased, the region over which the pressure drop occurs becomes narrower.

The effect of a differential-pumping aperture in front of the sample is that the sample experiences a lower pressure, defined by 2.5 in the molecular-flow regime. A sample $4R$ away from the aperture will experience a pressure of $0.985p_0$ on its surface at the location on the optical axis, which is a sufficiently small difference with respect to the background pressure for most experiments, including experiments using high-vapor pressure samples such as ice at temperatures close to the melting point. $4R$ can thus be considered as the upper limit for the optimal sample-aperture distance. As pointed out earlier, for $p_0 \approx 10$ Torr the effective length (i.e. $\sim 4R$) can only be a few hundred micrometers, requiring aperture diameters of the order of 100 μm or less.

To measure at higher pressures, the effective length has to be made smaller, by positioning the sample closer to the aperture. In this case, the aperture needs to be smaller in order to maintain $\sim p_0$ at the sample surface, and also to limit the gas flow to the first pumping stage. The greatest technical challenge in this situation is to focus the X-ray beam onto the sample area in front of the aperture, without the beam being obstructed by the aperture itself. Synchrotron sources have the advantage that they provide more tightly-focused beams with high flux and smaller divergence, as compared to lab sources. Even in that case, the size and shape of the front aperture have to be engineered carefully to meet the geometrical constraints given by X-ray beam size, divergence, and relative angle with respect to the optical axis of the spectrometer.

The flow of gas through the entrance and subsequent apertures and its effect on the pressure in the various differential-pumping stages can be estimated. In the molecular-flow case, the conductance of an aperture is calculated from the average number of collisions of the gas molecules with the walls, and the area of the aperture:

$$C = JA = \sqrt{\frac{kT}{2\pi m}}A, \tag{2.7}$$

with C as volumetric conductance, A as area of the aperture, k the Boltzmann constant, T the gas temperature, and m as the mass of a molecule of gas. For water vapor passing through a 100-μm aperture at room temperature (T = 298 K, $m = 3.0 \times 10^{-26}$ kg, A = 0.031 mm^2), the conductance is 5×10^{-3} L/s. As conservation-of-mass equations dictate $C_0 p_0 = C_1 p_1$, for a typical pumping speed of $C_1 \cong 100$ L/s behind the aperture, the pressure ratio between the volumes adjacent to the aperture, p_1/p_0, is 5×10^{-5}. The ratio can be expected to be larger for viscous or Knudsen-flow regimes, but not by much. A typical 4 orders of magnitude pressure drop is routinely attained with the current instruments and apertures of 100-μm diameter. A pressure of ~10 Torr at the sample translates to ~10^{-3} Torr in the first pumping stage, which is about the limit of molecular-flow regime, thus the operating limit of a turbomolecular pump. The apertures that separate the differentially-pumped volumes are typically larger than the first aperture (~few millimeters) and provide ~2 orders of magnitude pressure drop each.

In the preceding paragraphs we have defined a maximum pressure in the sample chamber $(p_0^{\max})$ at which APXPS can be performed efficiently. The goal of current technical developments is to increase $p_0^{\max}$ to allow for measurements under more relevant conditions in e.g. heterogeneous catalysis. The following analysis shows that an increase in this maximum sample chamber pressure limit does not lead to higher pressures in the differential pumping stages. The practical electron-scattering limit dictates that the product of the optimal sample-aperture distance and maximum pressure on the sample should be approximately constant, since the optimal sample-aperture distance is proportional to the aperture radius:

$$Rp_0^{max} \cong constant \tag{2.8}$$

The pressure in the first pumping stage is proportional to the pressure on the sample and the area of the aperture (i.e. $4\pi R^2$).

$$p_1 \propto R^2 p_0^{max} \tag{2.9}$$

Substituting (2.8) in (2.9) gives:

$$p_1 \propto \frac{\left(Rp_0^{max}\right)^2}{p_0^{max}} \tag{2.10}$$

$$p_1 \propto \frac{1}{p_0^{max}} \tag{2.11}$$

This means that a higher maximum sample-chamber-pressure limit does not translate to a higher pressure in the first pumping stage since the effect of the mandatory decrease in the aperture diameter will outweigh the pressure increase. The same argument may also be used for the latter pumping stages. The flux of the portion of the gas that travels through the first aperture close to the optical axis without being scattered (and thus pumped), eventually reaching the electron-energy analyzer is also proportional to the conductance of the first aperture, thus proportional to R^2. The conclusion is that the pressure limit in current APXPS instruments can be increased as long as the diameter of the first aperture is decreased proportionally.

2.1.2.3 Differentially-Pumped Electrostatic-Lens Systems

The cost of introducing a differential-pumping stage (apart from the financial "cost") is decreasing the collection angle of electrons to a solid angle set by the aperture that subtends the smallest solid angle as seen from the sample (see Fig. 2.3a). Just considering the influence of the first aperture on the detection efficiency, for a sample $4R$ away from the first aperture, the acceptance angle is limited to $2\arctan(R/4R) = 28°$. In the absence of electrostatic lenses, apertures in the latter pumping stages would cut down the acceptance angle dramatically. For example, an aperture of 3 mm diameter, 1 m away from the sample subtends an angle of 0.34°, as compared to the 28° limit of the first aperture, and it would cut down the electron count by almost 4 orders of magnitude. This reduction of electron transmission by a series of apertures in the differentially-pumped lens system was the limiting factor in the first-generation APXPS instruments, all of which were also using laboratory X-ray sources [16, 17].

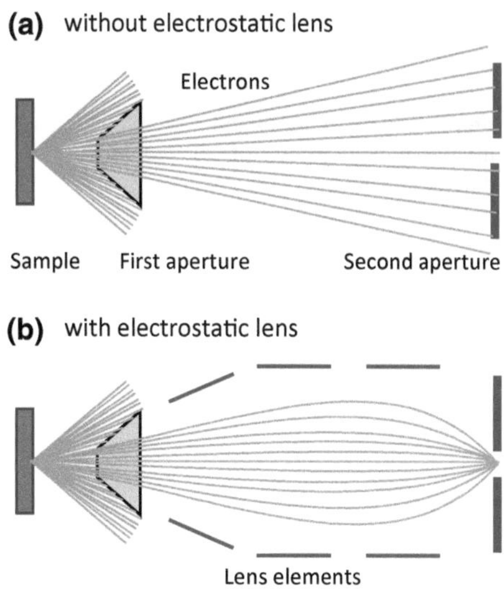

Fig. 2.3 Schematic illustration of the increase in the electron-collection efficiency with the use of an electrostatic-lens system. An electrostatic-lens system in the first pumping stage is close enough to the sample to collect electrons from a full-angle of ~30°

The electrostatic-lens system in an XPS instrument serves several purposes, among them to increase the efficiency with which the electrons are collected from the sample, and it matches the kinetic energy of the electrons to the pass energy of the analyzer. In many recent versions of ambient-pressure XPS instruments, electrostatic lenses are placed in between the differentially-pumped apertures, which increases the acceptance angle markedly, comparable to instruments with conventional electrostatic-lens systems without differential pumping (see Fig. 2.3b).

2.1.2.4 X-ray Sources for Ambient-Pressure XPS

X-ray sources, such as a conventional X-ray tube or a synchrotron, operate under UHV, and for that reason the high vacuum in the X-ray source has to be separated from the ambient pressure in the measurement cell. X-rays are much more weakly absorbed/scattered by matter than electrons. Thus, X-rays can pass through much thicker membranes (i.e. X-ray windows), which separate UHV and high pressure, than electrons. X-rays are also less attenuated by a gas than electrons.

Al, Si and silicon nitride (Si_3N_4) are being used as X-ray-window materials. While thin Al membranes have the highest transmission across the soft X-ray regime, they often develop pinholes and thus vacuum leaks. Si_3N_4 has proven to be the most effective and is the most widespread choice for X-ray windows in APXPS experiments. This is mostly due to the commercially developed, robust, and cheap Si_3N_4 windows. X-ray windows typically have sub-micron thicknesses. Transmission of soft X-rays through Si_3N_4 windows of different thicknesses is shown in Fig. 2.4. It can be seen that typically used, 50–100 nm-thick windows have more than 50% transmission above 300 eV, which is the useful energy range for most XPS experiments. The mechanical stability of Si_3N_4 windows allows for pressure differentials of 1 atm across an active window area of $0.5 \times 0.5\ mm^2$ for 100 nm-thick membranes.

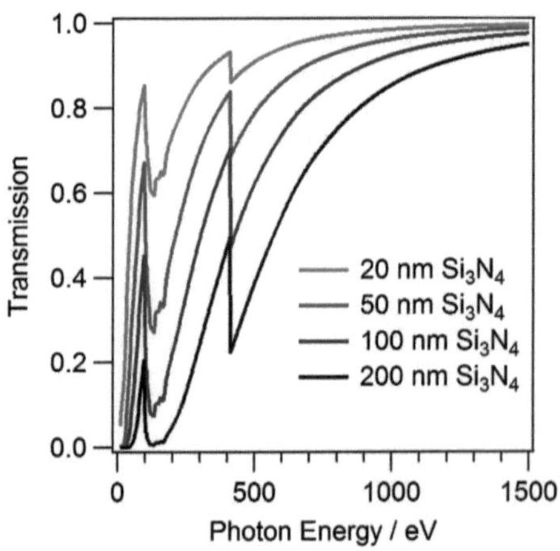

Fig. 2.4 X-ray transmission through Si_3N_4 windows of different thicknesses [18, 19]

Due to the tight focus and small divergence of X-rays produced by synchrotrons the active X-ray window size can be kept very small in those experiments, which allows the use of thinner membranes (typically 100 nm). For lab-based APXPS experiments monochromatized and focused X-ray sources are preferable (yet much more expensive than unfocused sources), but have a much larger divergence than X-rays originating from a synchrotron source. This requires to either use much larger active window areas or to place the window very close to the sample, which can be problematic due to space constraints in the vicinity of the sample and also the risk of exposing the window to high temperatures during experiments that involve heating the sample. Since laboratory-based X-ray sources in general produce X-rays with energies well in excess of 1 keV, where the transmission of Si_3N_4 is high, thicker and thus larger membranes can be used which can in turn be placed farther away from the sample's surface without cutting into the incident X-ray beam.

2.1.3 Phenomena that Are Relevant to XPS Analysis of Catalysts Under Gas Ambient

2.1.3.1 Monitoring the Gas Phase Over a Solid

When a solid (or liquid) sample is analyzed with APXPS, X-rays not only ionize the sample but also ionize the gas that they traverse. Some part of this ionized gas always falls into the electron-collection volume of the analyzer. This is illustrated in Fig. 2.5a. As a result, peaks of the gas molecules also show up in the XP spectra when the gas partial pressure is above ~0.1 Torr. Photoelectron peaks of gas molecules show up at a few eV higher binding energies than their condensed phase counterparts, and usually don't overlap with them. This constitutes a great way to analyze the gas composition right above the sample simultaneously with the surface composition, and is especially useful when studying molecular transformations, as in the case of catalysis.

As already mentioned, the sample is normally at the same potential as the analyzer (in the case of metals), so their Fermi levels are aligned and a change in the sample's work function does not affect the measured binding energies of electrons from the sample. On the other hand, the reference point of binding energy in gases is the vacuum level. As a result, the observed binding energies from the gas molecules change with the vacuum level of the volume of the gas that is ionized. This is illustrated in Fig. 2.5b where E_{vac}^{gas} is shown to be a linear function of E_{vac}^{sam} (sample) and E_{vac}^{ap} (aperture). Since $E_{vac} = E_F + \Phi$ and all the Fermi levels are aligned, changes in the work function of the sample and the aperture surfaces will cause changes in the observed binding energies for gas molecules.

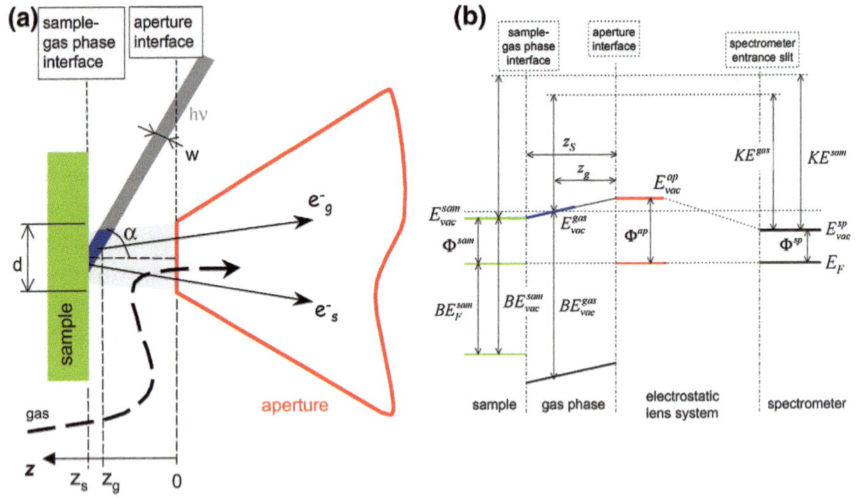

Fig. 2.5 **a** The typical sample/aperture/X-ray beam configuration in an APXPS setup. Part of the gas that is illuminated by the beam is in the acceptance cone of the spectrometer (i.e. blue area). **b** Energy level diagram corresponding to the region in part (**a**). Fermi levels are aligned due to the sample and the whole spectrometer being in electrical contact. The work functions of different parts of the setup are in general different, which results in different vacuum levels. The vacuum level of the gas that is seen by the spectrometer is a function of position, and changes linearly between the sample and the first aperture. Reproduced with permission from [12]

In addition to being affected by the changes in the work functions of the nearby surfaces, putting a bias on the sample (i.e. shifting the Fermi level) will also induce a shift in the gas peaks. This shift will be smaller than the bias applied, and its magnitude will be:

$$\Delta BE^{gas}_{vac} = -\Delta E^{gas}_{vac} = \left(z_g/z_s\right) eV_{bias} \tag{2.12}$$

where e is the magnitude of the electron's charge (positive in sign) and V_{bias} is the voltage applied to the sample. Keep in mind that the change of the binding energies from the sample will be eV_{bias}. This implies that overlapping gas and sample peaks can be separated by applying a high-enough bias to the sample.

In addition to monitoring the gas phase over a solid sample, APXPS can also be used to analyze the gas phase exclusively. Apart from obtaining fundamental electronic-structure information about the molecule, this can provide valuable quantitative information. Intensities from core levels from a material with ideal stoichiometry (such as a gas molecule) can serve as an almost-perfect internal reference. Using such an approach provides a correction factor for the photon flux, and the analyzer transmission function at those kinetic energies. The accuracy of the approach is fundamentally limited by the error that may be introduced by intramolecular electron-scattering effects [20].

A common case for this kind of quantitative analysis is the deconvolution of the O 1s signals of hydroxyls and carbonates on a solid surface, which have almost identical O 1s binding energies. Based on the C 1s and O 1s spectra from a C and O containing gas (e.g. CO_2, CO) as reference, one can calculate an O 1s/C 1s sensitivity factor. Using the C 1s signal of the carbonate and the sensitivity factor, the contribution of the carbonate to the O 1s intensity at the hydroxyl region can be found.

2.1.3.2 Beam Damage

Irradiation by X-rays can cause chemical changes in materials, mostly due to the generation of low-energy (secondary) electrons. The amount of damage depends on the X-ray flux (i.e. photons per area), the energy of the incident X-rays, as well as the chemical composition of the sample. Examples for beam-induced chemical changes are the reduction of oxides and the loss of anions in alkali halides upon irradiation. Conventional anode sources have a relatively low X-ray flux density compared to synchrotron sources and thus beam damage is expected to be slower in the former case. When beam damage takes place on a time scale comparable to that required to measure a spectrum it is easy to check for it by observing changes to the APXPS spectra with time. However, one should always keep in mind that beam damage can happen on a time scale not easily detectable in an XPS experiment (i.e. on a sub-second time scale) which is thus difficult to guard against, unless an independent verification method for the absence of beam damage is available, which to our knowledge has not been implemented so far in an APXPS experiment. Simultaneous IR and APXPS measurements with full spatial overlap might be a route to achieve control over beam damage. Although there are plans for implementing combined vibrational and core-level ambient-pressure spectroscopy experiments in a number of laboratories, a setup that would allow for these measurements has not yet been commissioned.

One way of avoiding beam damage is to continuously prepare a fresh sample surface, as in using a continuous stream of the sample, where the sample is replenished at a rate greater than the rate of beam damage. For a liquid sample, this might be a liquid jet or a droplet train [21–23]; for a solid, a jet of aerosol (solid/gas) or suspension (solid/liquid) [24].

2.1.3.3 Charging

Charging is a major problem in XPS when working with non-conducting samples such as catalysts supported on high band gap materials, e.g. alumina or silica. Charging not only leads to shifting of peaks (in some cases by hundreds of eV) but also to asymmetric broadening in case of inhomogeneous charging. While rigid peak shifts due to homogeneous charging across the whole investigated sample surface can be corrected for by using internal binding-energy standards, asymmetric broadening makes even relative binding-energy assignments impossible. In UHV, working with charging samples is possible by using an electron flood gun. In this case, low-energy electrons flood the sample and compensate the positive charge accumulated at the sample surface due to the emission of photoelectrons. The flood gun contains a hot emitter and cannot operate above a certain pressure ($\sim 10^{-8}$–10^{-6} Torr depending on the type of emitter). A differentially-pumped flood gun could be the solution but it has not been realized yet.

At a sufficiently high gas pressure in the reaction cell, electrons emitted from the gas molecules hit the sample and partially or completely compensate the charging of the sample. These are mostly secondary electrons but Auger and photoelectrons contribute as well. The extent of compensation depends on many factors such as the chemical composition of the sample/gas, photon flux, photon energy, X-ray incidence angle, and gas pressure. This effect may remove charging completely; but even when it does not the spatial inhomogeneity of the charging may be reduced to an extent that relative binding energies can be measured. In that case a peak with a well-known binding energy can be used as an internal reference to calibrate the binding energy scale.

Figure 2.6 shows the apparent Si 2p binding energy of a muscovite mica sample as a function of water vapor pressure. Muscovite mica is a layered aluminosilicate with intercalated K^+ ions, and its electrical conductivity is similar to aluminas, silicas, or aluminosilicates used as catalysts or catalyst supports. The nominal Si 2p binding energy in mica ($\sim$102.4 eV), measured in UHV using a flood gun is shown by a horizontal dotted line [12]. The shift of the Si 2p peak in 0.02 Torr water vapor is almost 150 eV, which is not uncommon, and the peak shapes at high charging conditions are quite asymmetric (see the inset spectra in the figure), making interpretation of chemical shifts impossible. As the water vapor pressure is increased from 0.02 to 1.75 Torr, the compensation of charge increases and the apparent binding energy gets closer to its nominal value, not reaching it exactly though. The broadening of the Si 2p peak due to inhomogeneous charging disappears almost completely at 1.75 Torr. In that regard, the increased mobility of K^+ ions under humid conditions, where a thin layer of liquid water forms on the surface, provide another mechanism for distribution of charge uniformly on the surface.

2.1.3.4 Heating and Cooling of Samples

One of the main goals in the field of catalysis research is to find catalysts that perform well at low temperatures, which would decrease the energy costs in industrial processes. Nonetheless catalysis, for the most part, is still a high-temperature science, where below-room-temperature reactions are rare [25]. When it comes to fundamental studies though, temperature may need to be manipulated in either way.

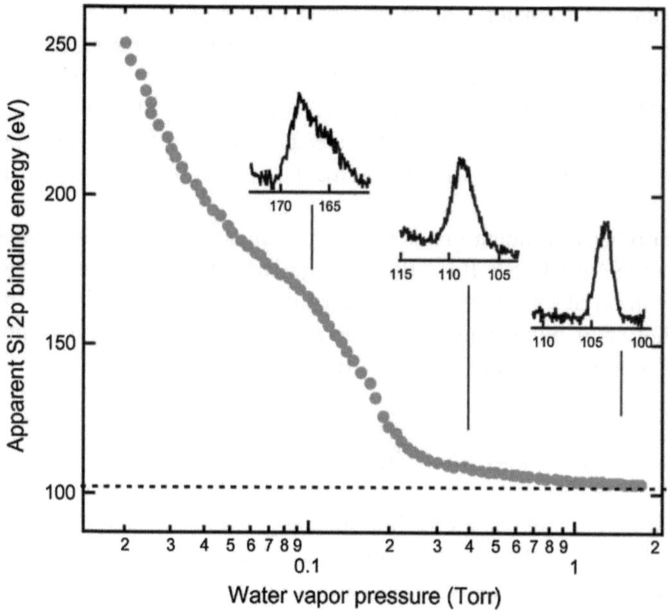

Fig. 2.6 Apparent Si 2p binding energy of a muscovite-mica sample as a function of water vapor pressure. The incident photon energy is 390 eV, with a flux density of 3×10^{11} mm^{-2} s^{-1}. The dotted line indicates the nominal Si 2p BE in muscovite mica. The insets show Si 2p spectra at water vapor pressures of 0.1, 0.4, and 1.7 Torr. Differential charging, which is observed at the lower pressures, is not observed at 1.7 Torr. Reproduced with permission from [12]

Cooling of samples for APXPS experiments constitutes a challenge. Regular schemes used for UHV-type experiments usually form the coldest spot outside the sample, such as on cooling rods, tubes, etc. In that case significant condensation/deposition may take place on these surfaces if the coldest spot has a lower temperature than the condensation point of the gas in the chamber. If the gas is introduced at a smaller rate than the rate of condensation, the pressure in the chamber would be limited by the temperature of the cooling element. In order to prevent such problems, samples can be cooled with Peltier elements, which ensure that the sample is the coldest point in the experimental setup [2].

Heating samples to study catalytic reactions also has some restrictions. Any kind of heating technique that uses a hot filament should use a chemically-resistant filament material, such as Pt. Common materials like W, Ta, and Ir will readily oxidize in oxidizing gases and would be of limited use. However, heating techniques that employ a hot metal such as a Pt filament are inherently problematic if one wants to study a reaction catalyzed also by the heater material. Resistive heaters where the metal is embedded inside a non-reactive ceramic material help overcome this problem.

One example of resistive heating has been applied at the Boreskov Institute in Novosibirsk, Russia [26, 27]. The instrument is a commercial VG-ESCALAB high-pressure spectrometer. The most recent design of the sample holder uses W wires spot-welded to a single-crystal sample for resistive heating [27]. The small area of the heater wire combined with the relatively inert nature of W surfaces makes the heater applicable for inert or reducing gas atmospheres. A special, pyrolytic boron nitride (PBN)-coated heater produced by Sintec Keramik has also been used in combination with the same instrument [26], allowing the use of oxidizing gases.

Another other instrument that is specially designed for catalysis studies is at the ISISS beamline at BESSY II, in Berlin [28]. That instrument uses an IR laser for heating, with an absorbing contact plate behind the sample. This heating scheme prevents hot spots other than the sample and has been successfully applied for a number of studies involving several different reactions [29–33].

2.1.3.5 Introducing and Analyzing Gases

There are different approaches for introducing gases into an APXPS chamber. The simplest one involves the use of variable leak valves under full, partial, or minimum pumping conditions, pumping being carried out by a turbomolecular pump. Full pumping utilizes the maximum capacity of the turbo pump; thus the pressure in the chamber has to be below $\sim 10^{-3}$ Torr. With a catalyst surface under investigation, this case is similar to a continuous stirred-tank reactor (CSTR) with very high space velocity. At this pressure range, mass transfer by diffusion is faster than the reaction rate and a CSTR approximation is valid. Partial pumping can be achieved by decreasing the conductance to the pump using a valve. Partial pumping increases the pressure in the chamber and decreases the space velocity. The mass-transfer rate decreases due to increased pressure and mass-transfer limitations start becoming significant. Minimum pumping is the case where the chamber is pumped only via the first aperture of the differential-pumping system. Because of the small conductance of the aperture, the space velocity is very small and the system is nearly a batch reactor. The limitation of leak valves is that it is difficult to set exact partial pressures for gases in a gas mixture. This problem can be partially overcome if the gas mixture in the chamber is analyzed by means of a mass spectrometer. It would still not be possible to set the partial pressures precisely but at least to measure them.

The second approach to introducing gas mixtures is using mass flow controllers (MFC). MFCs provide control on the flow rate of each individual gas and thus allow the user to set the desired composition of the inlet gas mixture. The pressure range that one can have using MFCs depends on the flow-rate limits of the MFC and the specifications of the pump used to pump the chamber. Working in the Torr range is possible as well as in high vacuum. MFCs are ideal for experiments in the Torr range with high space velocity.

2.1.3.6 Contamination

Under pressures relevant for APXPS (mTorr to Torr range), a surface site experiences 10^4–10^6 collisions per second. This means even gas impurities with ppm concentration may cover the surface in seconds if the sticking coefficients are not too low. Furthermore, high impingement rates on the chamber walls may also lead to additional impurities. Molecules on the walls may be simply displaced or react with the gas to form new contaminants, eventually ending up on the sample surface. Baseline pressure is not a sufficient indicator of the severity of possible contamination. Even when the baseline pressure is low ($<10^{-9}$ Torr), the walls may have a significant coverage of impurities. The nature of these impurities depends on the history of the chamber.

An example of contamination from a small impurity is Ni carbonyls that form in Ni-containing steel gas lines under high CO pressures and which can decompose on surfaces (especially at high temperatures) to leave metallic Ni. The contamination can be prevented by using Cu tubing and Ni-carbonyl traps. An example for impurities that form due to wall reactions is the formation of elemental halogens by the reaction of oxidizing gases with halide ions on the walls. In our group, we have observed fluorine, chlorine, bromine, and iodine contamination on our samples after introducing NO_2 to our chambers in the mTorr range. O_2 sometimes causes a fluorine contamination which we believe to happen via the same mechanism. Needless to say, this kind of contamination can be prevented only by keeping the chamber walls clean. In certain cases creating a nitrogen plasma in the chamber (followed by a bake-out) may help reduce wall contamination. The best approach to avoid impurities would be to use exchangeable cells, each of which is to be used for one reaction exclusively. Such cells are currently commercially available and are expected to be more widely used for studies in heterogeneous catalysis. Another type of contamination that is relevant at high pressures is the cross contamination that occurs while co-dosing the vapor of a liquid "A" and another gas/vapor "B" that is soluble in that liquid. If the pressure in the chamber and the pressure above liquid A is not high enough, the diffusion rate of B through the dosing valve into vapor/liquid A can be significant. An example is the diffusion of ammonia (NH_3) gas into liquid water during co-dosing. Liquid water can hold as much NH_3 as to end up in the vapor phase for the next experiment. Replacing the contaminated water with fresh water will solve the problem in this case.

2.1.3.7 Reactor Modeling

Industrial reactors may sometimes have limited reaction rates due to mass transfer from the bulk of the reactant stream to the catalyst surface and this fact is always taken into consideration in the reactor's design stage. Catalysis studies under realistic pressures also in general account for this fact. Surface-science studies of model catalysts on the other hand never had to take this into consideration, because the mean free path of molecules at or below high vacuum is larger than the reactor cell dimensions, and the number of reactant molecules striking the surface is low so that the rate of mixing of gas molecules is always much larger than the rate of reaction. Since APXPS experiments can be performed at pressures where mass transfer becomes an issue, a meaningful interpretation of the data necessitates the understanding of the concentration and temperature gradients in the system.

In an illustrative study, Blomberg et al. reported on real-time monitoring of CO and CO_2 concentrations over a Pd(110) single crystal during catalytic CO oxidation [34]. Their experimental chamber and the CO concentration map (obtained by laser-induced fluorescence) over the Pd(110) surface at nearly steady-state is shown in Fig. 2.7. The reaction was carried out under flow conditions (space velocity: ~1 s^{-1}) at 80 Torr total pressure and a CO:O_2:Ar reactant mixture of 1:1:2 ratio. The concentration map for CO shows a CO-depleted region above the Pd(110) surface. This region extends ~5 mm in every direction from the surface.

The conditions in APXPS experiments are usually different from the conditions studied by Blomberg et al. in the sense that flow rates of gases are smaller and the pressures involved are typically 2-3 orders of magnitude lower. In addition, the presence of the nozzle near the sample affects the concentration gradients.

One of the tools to incorporate mass (and heat) transfer effects into the analysis of APXPS data is multiscale modeling of experimental chambers. An example of this, although not an APXPS setup, is the modeling of the chamber shown in Fig. 2.7a [35]. In that work, Matera et al. show that it is possible to extract information about the state of the catalyst surface by combining Monte Carlo simulations (energies computed with DFT) and computational fluid dynamics. Including similar simulations to APXPS experiments in the future can surely contribute to data interpretation.

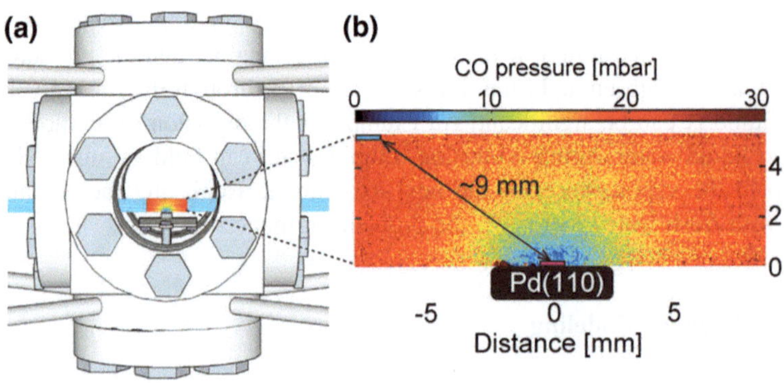

Fig. 2.7 Monitoring the concentration gradient in the gas phase during CO oxidation over a Pd (110) crystal. The total pressure is 80 Torr, with a CO:O_2:Ar ratio of 1:1:2. **a** Reactor setup showing the sample and the laser sheet used to monitor the CO concentration with 2D spatial resolution. **b** The CO partial pressure map above the sample measured via laser-induced fluorescence. Reproduced from [34] under the "Standard ACS Author Choice/Editors' Choice Usage Agreement"

2.2 Examples

2.2.1 CO Oxidation on Ru

There has been a hot debate about the active phase(s) of Ru catalysts in the CO oxidation reaction [36–38]. It was known for a long time that the Ru(0001) surface is inactive for CO oxidation under UHV conditions [39–41], however, it is remarkably active in the Torr pressure range [42, 43]. Under these conditions, Goodman and Peden proposed a O-(1 × 1) structure on Ru(0001) (~1ML of O), whereas Over and co-workers argued for a $RuO_2(110)$ surface, which forms under reaction conditions as the catalytically-active phase. Several ex situ, in situ, and *operando* techniques have been used to address the question. APXPS was used to investigate the reaction on Ru(0001) [44] and Ru(10–10) [45] single crystals, nanoparticles [46], and thin films [47]. The current understanding about this system is that even a poorly-ordered oxide layer (1–3 ML of oxygen) on Ru metal is active in CO oxidation, i.e. an ordered surface such as $RuO_2(110)$ is not necessary [48].

Blume et al. studied CO oxidation under different CO/O_2 ratios over a pressure range of 10^{-4}–10^{-1} Torr [44]. The sample was a Ru(0001) single crystal, which showed different oxides as a function of reaction conditions. The reaction product (CO_2) was monitored using a mass spectrometer that was mounted at the first differential-pumping stage of the electron spectrometer. O 1s and Ru $3d_{5/2}$ spectra were monitored in a 0.075 Torr, 1:1 $CO:O_2$ mixture, while the temperature was increased at a constant rate. Changes in the O 1s spectrum indicated a transition at ~500 K from a surface oxide, where the top few layers are incorporated with ~1–3 ML of O atoms, to a rutile-RuO_2 phase, which is expected to have the $RuO_2(110)$ orientation [49]. This transition, however, did not correspond to a change in the CO_2 production rate. It was concluded that "there was no distinct difference in catalytic activity between the stoichiometric $RuO_2(110)$ and a few-layers-thick, poorly-ordered surface oxide".

Toyoshima and coworkers studied CO oxidation on Ru(10–10) [45]. The reaction was carried out in a 0.11 Torr gas mixture with 1:10 $CO:O_2$ ratio, under flow conditions, at temperatures ranging from 390 to 570 K. The formation of CO_2 was monitored using a mass spectrometer installed at the differential-pumping system of the electron spectrometer. Transformation of the surface around 460 K from a chemisorbed oxygen state to a ~ 1 nm thick RuO_2 layer (equivalent to 4–5 atomic layers) was observed with APXPS. At the temperatures investigated, oxide growth is kinetically limited. RuO_2 is expected to have the (100) orientation [50]. The transformation was correlated with an increase in reaction rate, observed by an increase in CO_2 mass signal, and it was concluded that the thin RuO_2 layer was the catalytically active phase for the reaction.

Qadir et al. studied the surface structure of Ru catalysts under CO, O_2, and a CO + O_2 mixture (0.08 Torr CO + 0.2 Torr O_2) [46, 47]. The difference in catalytic activity of 2.8 and 6 nm-sized Ru nanoparticles, which had been reporter earlier [51], was attributed to the different thicknesses of the oxide layers that grew on the nanoparticles under reaction conditions [46]. Particles having a 2.8 nm diameter grew a thicker oxide layer than those of 6 nm diameter, which in turn suppressed the catalytic activity. The same type of deactivation behavior was also observed for Ru in the form of a thin film [47]. Thin-film Ru formed a thicker oxide under oxidizing conditions and this oxide layer was more difficult to reduce than the oxide layers that formed on nanoparticles.

2.2.2 CO Oxidation on Pd

Unlike Ru, metallic surfaces of Pd, Pt, and Rh are very active for CO oxidation. While the metallic surface is probably the most active state in CO oxidation [39, 52, 53], APXPS studies showed that surface oxides that form on these metals under reaction conditions may have significant activity too.

Blomberg et al. studied CO oxidation over Pd(100) using APXPS, up to a total pressure of 1 Torr, with 1:1 and 1:4 CO:O_2 ratios and over a temperature range of 420–680 K [54]. In a 1:1 CO:O_2 gas ambient, they observed the transformation of a CO-poisoned surface to an oxygen-covered metallic surface at increasing temperatures with increasing total pressure (e.g. 523 K for 0.01 Torr and 608 K for 0.5 Torr). The transformation was concomitant with the depletion of CO and the emergence of CO_2 in the gas-phase spectra, which implied that the new surface was catalytically active. Up to a total pressure of 0.5 Torr, the oxygen coverage of the active surface was below 0.4 ML, which is the range consistent with $p(2\times2)$ or $c(2\times2)$-overlayer structures. At a total pressure of 1 Torr, the oxygen coverage of the active surface increased significantly to 0.75 ML, consistent with a (5 × 5) oxidic-precursor structure observed earlier [55]. This increase in O coverage with pressure was interpreted as a sign of the pressure gap for this system existing somewhere above 1 Torr. With the use of a more oxidizing (1:4) CO:O_2 mixture at 0.5 Torr the authors observed the formation of a yet higher O coverage of 0.8 ML, which was interpreted as a $\sqrt{5}\times\sqrt{5}-R27$ surface-oxide phase [55, 56].

In a series of experiments Kondoh and co-workers studied CO oxidation over Pd (111), Pd(100), and Pd(110) [57–60]. Reaction products were monitored using a mass spectrometer placed at one of the differential-pumping stages. They used a CO:O_2 ratio of 1:10 (much more oxidizing than used by Blomberg et al.) with a total pressure of 0.22 Torr. Typical CO-poisoning behavior was observed at low temperatures (below 473 K, 463 K, and 438 K for Pd(111), Pd(100), and Pd(110), respectively), consistent with literature data. The use of an oxidizing gas mixture resulted in the formation of a $\sqrt{5}\times\sqrt{5}$ surface oxide on Pd(100) under reaction conditions (i.e. ~473 K). Figure 2.8 shows the CO_2 signal from the mass spectrometer as a function of the Pd(100) temperature, and Pd $3d_{5/2}$/C 1s spectra at certain points during heating, labeled A-D. Spectrum "E" corresponds to a bulk-oxide surface, which was not observed under reaction conditions but only under pure O_2 flow. The CO-poisoned region (A–B), the activation region (B–C), and the active region (C–D) can be seen from the CO_2 mass spectrometer signal in Fig. 2.8a. The activity of the surface oxide and the surface-oxygen content decrease with increasing temperature between points C and D. The authors concluded that an active oxygen species (i.e. "the up-side O" in the $\sqrt{5}$ surface oxide) was being lost at higher temperatures due to decomposition.

In CO-oxidation experiments using *operando* techniques such as APXPS, the emergence of a surface phase (such as an oxide) at the onset of catalytic activity is usually regarded as proof that this phase is responsible for the change in activity. Recent work by Matera et al. shows that this may not be the case for the well-studied Pd(100) surface [35]. By multiscale modeling of a catalytic reactor, they show that metallic Pd(100), which is a minority phase under the given CO-oxidation conditions, may be responsible for most of the observed activity. This work is a reminder that the activity of possible minority phases, which may not be quantified by XPS, can be responsible for the most, if not all, of the catalytic activity.

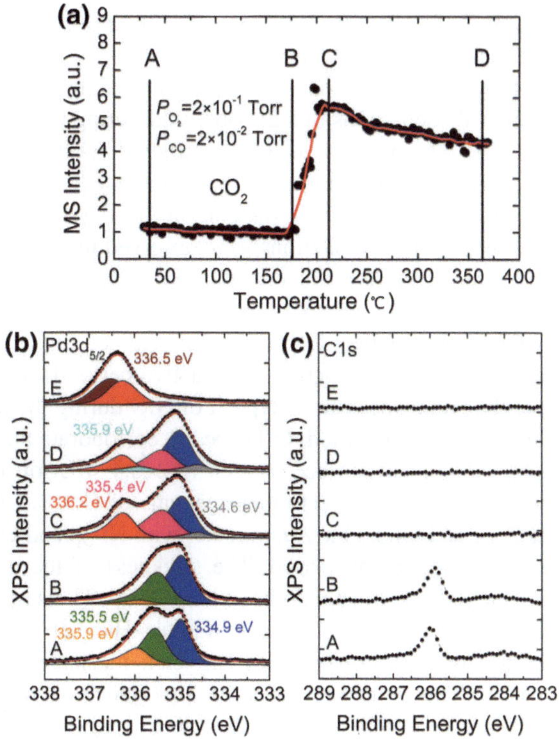

Fig. 2.8 CO oxidation on Pd (100) investigated by APXPS and mass spectrometry (MS). **a** CO_2 MS signal as a function of sample temperature under 2×10^{-1} Torr O_2 and 2×10^{-2} Torr CO. XP spectra are shown at points A–D. **b**, **c** Pd $3d_{5/2}$ and C 1s spectra, respectively, at points A–D and E. Spectra "E" were taken at 643 K, under 2×10^{-1} Torr pure O_2, after CO dosing was stopped. Adapted with permission from [58]. Copyright 2015 American Chemical Society

2.2.3 Catalytic Oxidation of Small Organic Molecules

Gabasch et al. studied the oxidation of CH_4 on the Pd(111) surface. A 1:5 reaction mixture of CH_4/O_2 was used at a pressure of 0.25 Torr [61]. The reaction products were monitored using two mass spectrometers, one in the first differential-pumping stage, and the other connected directly to the chamber via an adjustable capillary leak. The reaction shows a well-known activity hysteresis for the (111) surface as a function of temperature. The study looked into this effect by monitoring the surface composition and structure during reaction, both during heating and cooling. A detailed analysis involving rigorous peak fitting was used to identify the different oxygen species at the Pd(111) surface [62, 63]. It was observed that during heating, in the 530–650 K range, PdO seeds form on the 2D Pd_5O_4 surface oxide. The combination of these two oxides has a high catalytic activity, which translates into increased CO_2 and H_2O production during heating of the sample. During cooling, the increased catalytic activity was not observed, and neither was the Pd_5O_4 + PdO seed mixture. The surface oxide present during cooling had even less O than Pd_5O_4. The absence of PdO seed formation during cooling was explained on the basis of kinetic limitations to oxygen supersaturation, which is probably needed for the nucleation of the PdO seeds.

Kaichev et al. used APXPS to determine the formation mechanism of self-sustained oscillations that are observed during partial oxidation of propane over Ni foil [64]. The oscillations were monitored via the quantitative chemical analysis of the gas mixture using mass spectrometry and gas chromatography. The total pressure in the chamber was constant at 0.38 Torr. When the propane-to-oxygen ratio was varied, oscillations were observed for ratios of 3:1–15:1 (propane:oxygen) and over a temperature range of 873–973 K. Each period of the oscillation is divided into two regions, an active and an inactive half-period. The active region is typically shorter than the inactive region (see Fig. 2.9) and the ratio of the catalytic activity between the active and inactive period is around 40. The authors observed that the Ni 2p spectrum is typical of NiO during the inactive period and metallic Ni during the active period. O 1s spectra corroborate this finding while also showing a small amount of surface oxygen (0.2–0.4 ML) on metallic Ni. The sample temperature and mass spectrometer signals of CO, H_2, O_2, and H_2O as a function of time are shown in Fig. 2.9, together with Ni 2p and O 1s spectra before, during, and after the active half-period. The thickness of the oxide layer during the inactive half-period is estimated to be at least 3 nm. The authors of this study conclude that the reaction proceeds via a Langmuir–Hinshelwood mechanism on the metallic surface whereas a Mars–van Krevelen mechanism is at play when the oxide is present. This work illustrates the importance of in situ studies since these oscillations are very difficult or impossible to study using ex situ measurements.

Another example of APXPS studies in heterogeneous catalysis is the investigation of the selectivity of Ag-based catalysts for the ethylene-epoxidation reaction, which was studied by Rocha et al. [65], specifically the effect of Cl as a promoter. The sample was a commercial Ag nanopowder (<100 nm) pressed into a pellet.

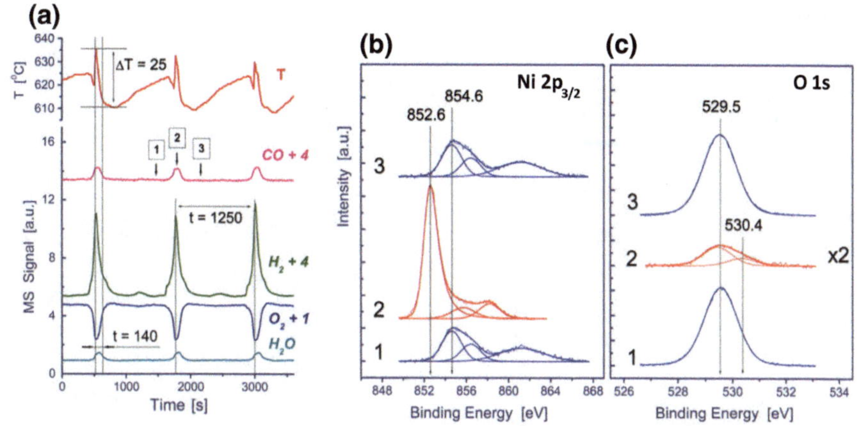

Fig. 2.9 Example for monitoring the oscillatory surface state of a catalyst surface (Ni foil) during partial oxidation of propane at 0.38 Torr with a propane:oxygen ratio of 8:1. **a** Catalyst temperature and MS signals of CO, H_2, O_2, and H_2O as a function of time (The MS signals from H_2, CO, and O_2 are vertically shifted by the factor shown in the legends). **b**, **c** Ni 2p and O 1s spectra before (*1*), during (*2*), and after (*3*) the short reactive half-period, respectively. Reproduced with permission from [64]

The chamber was kept at 0.23 Torr fixed pressure. An oxygen-rich feed with 1:2 $C_2H_4{:}O_2$ ratio was used with a flow rate of 4 mL/min. The gases were introduced using mass flow controllers. The sample temperature was fixed at 503 K using an infrared laser and a PID feedback. Gas chromatography was used to quantify the product concentrations and selectivity.

The analysis of the different oxygen species on Ag using APXPS is complex, and has been done thoroughly in earlier studies [66, 67]. Based on these references, the authors grouped different O 1s peaks under three categories: nucleophilic oxygen, electrophilic oxygen, and contamination (see Fig. 2.10b). The binding energies of the O 1s peaks increase in the same order. The electrophilic-to-nucleophilic oxygen ratio (O_{elec}/O_{nucl}) was used as a descriptor of the surface. During a 12-hour run on stream with an un-promoted sample, this ratio changes slowly over time and it correlates positively with the selectivity of the catalyst.

The results of the chlorine-promotion experiments are shown in Fig. 2.10. After more than 14 h on stream, Cl was added to the system by pulsing ethyl chloride (diluted in He) into the inlet stream (Fig. 2.10a). The surface of the Ag catalyst chlorinated immediately, as observed from the increase in the Cl 2p intensity (Fig. 2.10c). The selectivity of the catalyst increased, mostly due to a decrease in the CO_2 production (i.e. suppression of the combustion reaction). Introducing more Cl to the system with two more pulses resulted in more Cl on the surface and a further increase in selectivity. The correlation of O_{elec}/O_{nucl} ratio and ethylene oxide selectivity with chlorine content on the surface is clear (Fig. 2.10d). The increase in surface-Cl concentration resulted in a decrease of nucleophilic oxygen on the surface and concomitant decrease of the CO_2 production rate, whereas the amount

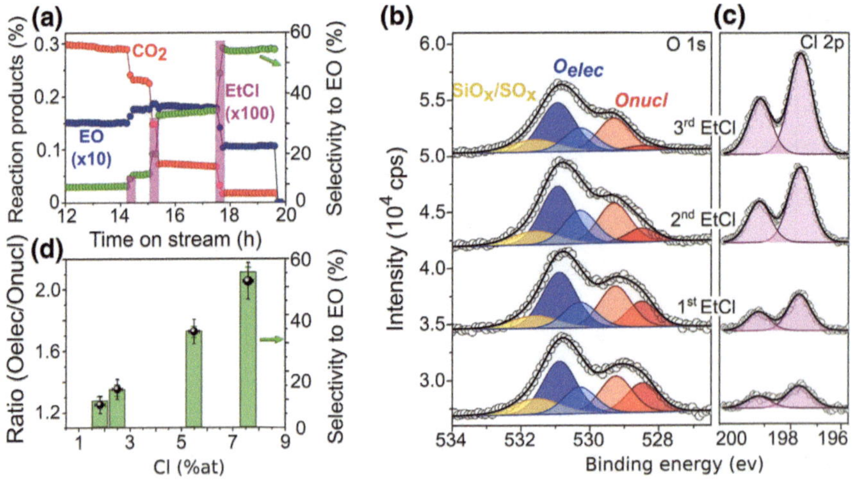

Fig. 2.10 Ethylene-epoxidation reaction studied using APXPS. **a** CO_2, ethylene oxide, and ethyl chloride concentrations in the chamber, and ethylene oxide selectivity as a function of time. **b**, **c** O 1s and Cl 2p spectra before and after ethyl chloride pulses. **d** correlation of O_{elec}/O_{nucl}, ethylene oxide selectivity, and chlorine concentration on the surface. Reproduced with permission from [66]

of electrophilic oxygen and the ethylene oxide production rate increased with the first two pulses but decreased on the third one. The total oxygen content of the surface and ethylene conversion also decreased after the third pulse. The results indicate that Cl promotes the selectivity via a site-blocking mechanism by mainly suppressing the combustion reaction through the removal of nucleophilic oxygen species from the surface. The increase in the concentration of electrophilic oxygen species was rather limited, but the increase in the ethylene oxide production rate was quantifiable. Excess Cl was detrimental for the activity of the catalyst.

2.3 Summary

We have discussed some of the main challenges for the use of XPS to *operando* studies of heterogeneous catalysts, and how they can be addressed using ambient-pressure XPS. The fundamental limit for increasing the pressure in APXPS studies is the scattering of electrons in the gas on their way from the sample surface to the entrance aperture of the differentially-pumped electron spectrometer. Currently, the pressures that one can work routinely is on the order of 10–20 Torr for water vapor, and less than that for molecules with stronger scattering cross sections, i.e. most gases. Expanding the pressure limit to more realistic conditions close to and above one atmosphere is a technical challenge that will most likely be overcome in the near future. Working at higher electron kinetic energies will be beneficial (although reducing the surface sensitivity). APXPS at higher electron kinetic energies (up to 10 keV) can enable the study of microporous and mesoporous catalysts like zeolites with higher sensitivity to the inner surfaces (e.g. pores) of realistic catalytic materials. The main avenue for increasing the pressure limit in APXPS is the reduction in entrance aperture size, which allows to reduce the path length of the electrons through the gas proportionally. Working at higher pressures will also help reduce charging of insulating samples, but at the same time mass-transfer effects have to be taken into account, in particular differences in the composition of the gas in the background of the chamber, the mass spectrometer, and the gas right above the surface. Modeling of fluid flow and concentration gradients will thus be an important part of data analysis for measurements at more realistic pressures.

Acknowledgements This work was supported by the Director, Office of Science, Office of Basic Energy Sciences, and by the Division of Chemical Sciences, Geosciences and Biosciences of the U.S. Department of Energy at LBNL under Contract No. DE-AC02-05CH11231.

References

1. A. Jablonski, C.J. Powell, J. Electron. Spectrosc. **199**, 27 (2015)
2. D.F. Ogletree, H. Bluhm, E.D. Hebenstreit, M. Salmeron, Nucl. Instrum. Methods A. **601**, 151 (2009)
3. S. Tanuma, C.J. Powell, D.R. Penn, Surf. Interface Anal. **21**, 165 (1994)
4. A. Kolmakov, D.A. Dikin, L.J. Cote, J.X. Huang, M.K. Abyaneh, M. Amati, L. Gregoratti, S. Gunther, M. Kiskinova, Nat. Nanotechnol. **6**, 651 (2011)
5. J. Kraus, R. Reichelt, S. Gunther, L. Gregoratti, M. Amati, M. Kiskinova, A. Yulaev, I. Vlassiouk, A. Kolmakov, Nanoscale **6**, 14394 (2014)
6. J.J. Velasco-Velez et al., Angew. Chem. Int. Ed. **54**, 14554 (2015)
7. C.H. Wu, R.S. Weatherup, M.B. Salmeron, Phys. Chem. Chem. Phys. **17**, 30229 (2015)
8. J.J. Velasco-Velez et al., Rev. Sci. Instrum. **87**, 53121 (2016)
9. R.S. Weatherup, B. Eren, Y. Hao, H. Bluhm, M.B. Salmeron, J. Phys. Chem. Lett. **7**, 1622 (2016)
10. M. Salmeron, R. Schlogl, Surf. Sci. Rep. **63**, 169 (2008)
11. D.F. Ogletree, H. Bluhm, G. Lebedev, C.S. Fadley, Z. Hussain, M. Salmeron, Rev. Sci. Instrum. **73**, 3872 (2002)
12. H. Bluhm, J. Electron. Spectrosc. **177**, 71 (2010)
13. M.E. Grass, P.G. Karlsson, F. Aksoy, M. Lundqvist, B. Wannberg, B. S. Mun, Z. Hussain, and Z. Liu, Rev. Sci. Instrum. **81** (2010)
14. D.R. Miller, in *Atomic and molecular beam methods*, ed. by G. Scoles (Oxford University Press, New York, 1988), p. 14
15. J.M. Kahk et al., J. Electron. Spectrosc. **205**, 57 (2015)
16. H. Siegbahn, K. Siegbahn, J. Electron. Spectrosc. Relat. Phenom. **2**, 319 (1973)
17. H. Siegbahn, J. Phys. Chem. **89**, 897 (1985)
18. B.L. Henke, E.M. Gullikson, J.C. Davis, Atom. Data Nucl. Data **54**, 181 (1993)
19. X-Ray Interactions With Matter, http://henke.lbl.gov/optical_constants/. Accessed 15 Jul 2015)
20. J. Soderstrom et al., Phys. Rev. Lett. **108** (2012)
21. D.E. Starr, E.K. Wong, D.R. Worsnop, K.R. Wilson, H. Bluhm, Phys. Chem. Chem. Phys. **10**, 3093 (2008)

22. M.A. Brown, R. D'Auria, I.F.W. Kuo, M.J. Krisch, D.E. Starr, H. Bluhm, D.J. Tobias, J.C. Hemminger, Phys. Chem. Chem. Phys. **10**, 4778 (2008)
23. M. Faubel, B. Steiner, J.P. Toennies, J. Chem. Phys. **106**, 9013 (1997)
24. M.A. Brown, I. Jordan, A.B. Redondo, A. Kleibert, H.J. Woerner, J.A. van Bokhoven, Surf. Sci. **610**, 1 (2013)
25. U. Dingerdissen, A. Martin, D. Herein, H.J. Wernicke, in *Handbook of Heterogeneous Catalysis* (Wiley-VCH Verlag GmbH & Co. KGaA, 2008)
26. V.I. Bukhtiyarov, V.V. Kaichev, I.P. Prosvirin, Top. Catal. **32**, 3 (2005)
27. A. Knop-Gericke et al., Adv. Catal. **52**, 213 (2009)
28. ISISS Station at BESSY II, https://www.helmholtz-berlin.de/pubbin/igama_output?modus=einzel&gid=1671. Accessed 28 Jul 2015
29. V.I. Bukhtiyarov, A.I. Nizovskii, H. Bluhm, M. Havecker, E. Kleimenov, A. Knop-Gericke, R. Schlogl, J. Catal. **238**, 260 (2006)
30. E. Kleimenov et al., Surf. Sci. **575**, 181 (2005)
31. D. Teschner et al., J. Catal. **242**, 26 (2006)
32. E. de Smit, F.M.F. de Groot, R. Blume, M. Havecker, A. Knop-Gericke, B.M. Weckhuysen, Phys. Chem. Chem. Phys. **12**, 667 (2010)
33. C. Rameshan et al., J. Catal. **276**, 101 (2010)
34. S. Blomberg, C. Brackmann, J. Gustafson, M. Alden, E. Lundgren, J. Zetterberg, ACS Catal. **5**, 2028 (2015)
35. S. Matera, S. Blomberg, M.J. Hoffmann, J. Zetterberg, J. Gustafson, E. Lundgren, K. Reuter, ACS Catal. **5**, 4514 (2015)
36. D.W. Goodman, C.H.F. Peden, M.S. Chen, Surf. Sci. **601**, 5663 (2007)
37. D.W. Goodman, C.H.F. Peden, M.S. Chen, Surf. Sci. **601**, L124 (2007)
38. H. Over, M. Muhler, A.P. Seitsonen, Surf. Sci. **601**, 5659 (2007)
39. T. Engel, G. Ertl, in *Advances in Catalysis*, edited by H. P. D.D. Eley, B.W. Paul (Academic Press, 1979), p. 1
40. H.I. Lee, J.M. White, J. Catal. **63**, 261 (1980)
41. V.I. Savchenko, G.K. Boreskov, A.V. Kalinkin, A.N. Salanov, Kinet Catal+ **24**, 983 (1983)
42. N.W. Cant, P.C. Hicks, B.S. Lennon, J. Catal. **54**, 372 (1978)
43. C.H.F. Peden, D.W. Goodman, J. Phys. Chem-us **90**, 1360 (1986)
44. R. Blume et al., J. Catal. **239**, 354 (2006)
45. R. Toyoshima, M. Shimura, M. Yoshida, Y. Monya, K. Suzuki, K. Amemiya, K. Mase, B.S. Mun, H. Kondoh, Surf. Sci. **621**, 128 (2014)
46. K. Qadir, S.H. Joo, B.S. Mun, D.R. Butcher, J.R. Renzas, F. Aksoy, Z. Liu, G.A. Somorjai, J. Y. Park, Nano Lett. **12**, 5761 (2012)
47. K. Qadir, S.M. Kim, H. Seo, B.S. Mun, F.A. Akgul, Z. Liu, J.Y. Park, J. Phys. Chem. C **117**, 13108 (2013)
48. H. Over, Chem. Rev. **112**, 3356 (2012)
49. H. Over, Y.D. Kim, A.P. Seitsonen, S. Wendt, E. Lundgren, M. Schmid, P. Varga, A. Morgante, G. Ertl, Science **287**, 1474 (2000)
50. Y.D. Kim, S. Schwegmann, A.P. Seitsonen, H. Over, J. Phys. Chem. B **105**, 2205 (2001)
51. S.H. Joo, J.Y. Park, J.R. Renzas, D.R. Butcher, W.Y. Huang, G.A. Somorjai, Nano Lett. **10**, 2709 (2010)
52. F. Gao, Y.L. Wang, D.W. Goodman, J. Phys. Chem. C **114**, 6874 (2010)
53. T. Engel, G. Ertl, J Chem Phys **69**, 1267 (1978)
54. S. Blomberg et al., Phys. Rev. Lett. **110** (2013)
55. M. Todorova et al., Surf. Sci. **541**, 101 (2003)
56. P. Kostelník, N. Seriani, G. Kresse, A. Mikkelsen, E. Lundgren, V. Blum, T. Šikola, P. Varga, M. Schmid, Surf. Sci. **601**, 1574 (2007)
57. R. Toyoshima et al., J. Phys. Chem. C **116**, 18691 (2012)
58. R. Toyoshima, M. Yoshida, Y. Monya, K. Suzuki, B.S. Mun, K. Amemiya, K. Mase, H. Kondoh, J. Phys. Chem. Lett. **3**, 3182 (2012)

59. R. Toyoshima, M. Yoshida, Y. Monya, K. Suzuki, K. Amemiya, K. Mase, B.S. Mun, H. Kondoh, J. Phys. Chem. C **117**, 20617 (2013)
60. H. Kondoh, R. Toyoshima, Y. Monya, M. Yoshida, K. Mase, K. Amemiya, B.S. Mun, Catal. Today (2015)
61. H. Gabasch et al., J. Phys. Chem. C **111**, 7957 (2007)
62. H. Gabasch et al., Surf. Sci. **600**, 2980 (2006)
63. D. Zemlyanov et al., Surf. Sci. **600**, 983 (2006)
64. V.V. Kaichev, A.Y. Gladky, I.P. Prosvirin, A.A. Saraev, M. Hävecker, A. Knop-Gericke, R. Schlögl, V.I. Bukhtiyarov, Surf. Sci. **609**, 113 (2013)
65. T.C.R. Rocha, M. Havecker, A. Knop-Gericke, R. Schlogl, J. Catal. **312**, 12 (2014)
66. V.I. Bukhtiyarov, M. Havecker, V.V. Kaichev, A. Knop-Gericke, R.W. Mayer, R. Schlogl, Phys. Rev. B **67** (2003)
67. T.C.R. Rocha, A. Oestereich, D.V. Demidov, M. Havecker, S. Zafeiratos, G. Weinberg, V.I. Bukhtiyarov, A. Knop-Gericke, R. Schlogl, Phys. Chem. Chem. Phys. **14**, 4554 (2012)

Chapter 3
Surface-Sensitive X-ray Diffraction Across the Pressure Gap

Andreas Stierle, Johan Gustafson and Edvin Lundgren

Abstract In this chapter surface-sensitive X-ray diffraction is introduced as an important crystallographic tool for the investigation of surfaces and nanostructures under high pressure reaction conditions and elevated temperatures which are relevant for industrial catalysis. After the introduction surface-sensitive X-ray diffraction methods are briefly explained and specialized instrumentation developed for the in situ investigation of surfaces and nanostructures across the pressure gap is presented combined with simultaneous measurement of the concentrations of reactants and products. In the following an overview of the experimental results is given: First the (near)- ambient pressure oxidation of 3d, 4d, and 5d transition metals is discussed which are relevant for oxidation catalysis. Afterwards catalytic reaction experiments in batch mode are reported, followed by an overview of current research using a flow reactor for surface-sensitive X-ray diffraction. Finally a perspective is given for future research directions.

3.1 Introduction

Surface-sensitive X-ray diffraction methods were introduced as a surface characterization tool in the mid-eighties of the last century, when hard X-ray synchrotron light from large particle accelerators became available for solid state physicists [1–3]. The method portfolio comprises surface X-ray diffraction (SXRD), grazing incidence X-ray diffraction (GIXRD), X-ray reflectometry (XRR), and grazing incidence small angle X-ray scattering (GISAXS). Since then, the methods became standard tools for the analysis of surface, interface and thin film structures. The

A. Stierle (✉)
Deutsches Elektronen Synchrotron DESY, Hamburg, Germany
e-mail: andreas.stierle@desy.de

J. Gustafson · E. Lundgren
Lund University, Lund, Sweden
e-mail: johan.gustafson@sljus.lu.se

E. Lundgren
e-mail: edvin.lundgren@sljus.lu.se

J. Frenken and I. Groot (eds.), Operando *Research in Heterogeneous Catalysis*,
Springer Series in Chemical Physics 114, DOI 10.1007/978-3-319-44439-0_3

main advantage of hard X-ray-based surface-sensitive diffraction techniques (photon energy 10–100 keV), as compared to electron-based surface characterization tools, is that they can be routinely applied in situ, under any type of surrounding environment at variable temperatures, especially during realistic gas pressures and compositions relevant for heterogeneous catalysis. In addition, the X-ray diffraction experiments are not hampered by insulating sample charging effects. Surface-sensitive X-ray diffraction methods are especially suited for the investigation of catalytic reactions, since they allow addressing the atomic structure of catalyst surfaces, which gives important insight into reaction mechanisms. This deeper level of fundamental insight into the surface structure of a working catalyst, as compared to X-ray powder diffraction or extended X-ray absorption fine structure (EXAFS) experiments, however, sets constraints on the sample itself: For all surface-sensitive X-ray diffraction methods a planar surface geometry is required; for crystallographic information single-crystal surfaces, epitaxial nanoparticles or nanostructures are mandatory. The wide range of temperatures and gas pressures accessible by surface-sensitive X-ray diffraction allows a direct comparison of the experimental results with theoretical calculations by *ab initio* density functional theory coupled to thermodynamics, as presented in Chap. 7 of this book.

This chapter is organized as follows: In the first part the concept of surface-sensitive X-ray diffraction methods is briefly introduced and novel sample environments for in situ experiments are described. In the second part, near-atmospheric pressure oxidation experiments of 4d and 5d metal surfaces and nanoparticles are discussed. The oxide phases are considered to be relevant as active phases for oxidation catalysis such as CO oxidation. In the third part, in situ studies under batch reaction conditions are discussed, which are complemented by *operando* studies under flow conditions (fourth part).

3.2 Surface-Sensitive X-ray Diffraction as in situ Tool

Surface-sensitive X-ray diffraction techniques can be divided in four different sub-methods, which allow to probe different properties of the surface. Surface X-ray diffraction (SXRD) provides crystallographic information on the atomic structure and composition of surfaces, interfaces, and nanoparticles [1, 3–5]. SXRD is complemented by X-ray reflectivity sensitive to the total electron-density profile perpendicular to the surface, containing information on roughness, layer thickness, nanoparticle average height, and surface coverage [6, 7]. In grazing incidence X-ray diffraction the information depth probed can be changed from nm to μm by a variation of the incident and the exit angle with respect to the surface [2, 8]. In addition, the signal-to-noise ratio in diffraction experiments can be optimized by a grazing incidence diffraction geometry. Finally, in grazing incidence small angle X-ray scattering (GISAXS) experiments the lateral surface morphology is probed [9].

Surface X-ray diffraction, grazing incidence X-ray diffraction, and grazing incidence small angle scattering experiments are traditionally performed using synchrotron radiation because of the required high brilliance of the X-ray beam to detect the diffraction signal from single atomic layers or very small nanostructures. Practically all third-generation synchrotron radiation sources possess end stations, which permit to perform SXRD, GIXRD and GISAXS experiments in combination with customized user sample environments [10]. X-ray reflectivity measurements are routinely performed using X-ray lab sources and with the advent of higher-brilliance micro-focus X-ray tubes also SXRD, GIXRD, and GISAXS experiments become feasible in the lab, at least for high-Z materials.

3.2.1 Basics of Surface X-ray Diffraction

The principles of surface X-ray diffraction (SXRD) are based on the fact that X-ray diffraction from a half-infinite crystal terminated by a surface (see Fig. 3.1a), is intrinsically sensitive to the atomistic structure of the surface. A real-space lattice made up by lattice vectors $\boldsymbol{a}$, $\boldsymbol{b}$, $\boldsymbol{c}$ is connected to the corresponding reciprocal space vectors $\boldsymbol{a}^*$, $\boldsymbol{b}^*$, $\boldsymbol{c}^*$ by the Laue equations [11]. A 3D infinite crystal gives (in kinematical approximation) rise to δ function-like Bragg reflections (blue circles in the reciprocal lattice in Fig. 3.1b). For a semi-infinite crystal, the Bragg reflections are not represented by δ functions any more and the so-called crystal truncation rods (CTRs) arise because of the truncation of the crystal by the surface. They run perpendicular to the crystal surface (HKL) plane through the Bragg reflections, as indicated by the green lines in Fig. 3.1b. In case the overlayer exhibits a different periodicity, additional surface rods arise (red dashed lines in Fig. 3.1b), which are sensitive to the structure of the overlayer only. In case of a commensurate structure the CTRs carry in addition information on the registry of the overlayer with respect to the substrate and the interfacial structure.

In a scattering experiment the scattering vector $\boldsymbol{q}$ is defined as $\boldsymbol{q} = \boldsymbol{k}_f - \boldsymbol{k}_i$, where the incident and exit wave vectors of the elastically scattered X-rays are given by $\boldsymbol{k}_{i,f}$, respectively ($|\boldsymbol{k}_{i,f}| = \frac{2\pi}{\lambda}$ with X-ray wavelength λ). To probe the reciprocal lattice, the Bragg condition $\boldsymbol{q} = \boldsymbol{G}$ has to be fulfilled, where $\boldsymbol{G}$ is a reciprocal lattice vector. The experiment can in principle be realized by any type of diffractometer possessing three independent degrees of freedom for the sample and detector stages. For practical reasons, most of the surface diffraction experiments are carried out on six-circle diffractometers in z-axis mode [1, 12]. Two out of the six circles are needed to orient the surface normal of the sample along the rotation axis θ, see Fig. 3.1c. The third axis, kept fixed during the experiments, is the incident angle μ. Typically, small incident angles are chosen, to reduce background scattering from the bulk and to increase the signal-to-background ratio. Note, that the surface sensitivity of the technique does not depend on the actual value of the incident angle—the reciprocal lattice of a two dimensional or half-infinite system is intrinsically surface sensitive.

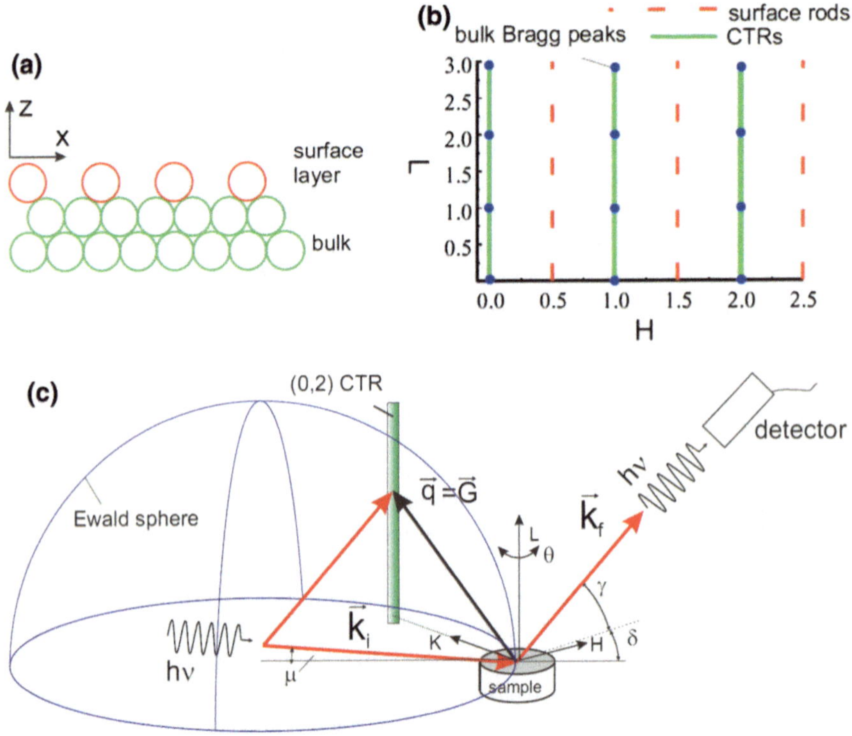

Fig. 3.1 **a** Truncated crystal with reconstructed adlayer. **b** Reciprocal lattice of the real space structure in (**a**). **c** *Z*-axis surface X-ray diffraction geometry

The intersection of the CTR with the Ewald sphere defines the direction of the diffracted X-rays (wave vector $\boldsymbol{k}_f$), which are collected by the detector rotating in δ and γ. For crystallographic structure-factor measurements, the diffractometer is moved to a specific (H, K, L) position and a θ scan is performed. The amplitude of the structure factor is obtained after integration of the θ scans and applying standard correction factors [5]. Alternatively, the structure factor can be collected using a two-dimensional detector [13, 14].

The diffracted CTR intensity can be calculated in a straightforward way [3]:

$$I(\boldsymbol{q}) \propto |F(\boldsymbol{q})|^2 \cdot \frac{1}{sin^2(\pi L)} \cdot \delta(\boldsymbol{Q}_{\|} - \boldsymbol{G}_{\|}) \tag{3.1}$$

$F(\boldsymbol{q})$ represents the structure factor of the unit cell, the second term is the CTR term, depending on the continuous reciprocal-lattice coordinate L in units of c*. It diverges at the positions of the bulk Bragg reflections at integer values of L. The third term is a δ function for the momentum transfer $\boldsymbol{Q}_{\|}$ parallel to the surface, because in the surface plane the Bragg condition $\boldsymbol{Q}_{\|} = \boldsymbol{G}_{\|}$ has to be fulfilled to observe scattered intensity ($G_{||}$ represents the component of the reciprocal lattice vector in the surface plane).

The following information can be obtained from a surface X-ray diffraction experiment by a trial-and-error fit to the data or by the use of so-called 'direct methods', based on phase-retrieval algorithms [15]: atomic positions, surface structure with pm resolution, surface composition, surface and interface roughness, thermal vibration amplitudes, and the overlayer registry to the substrate. In general, all additional experimental and theoretical information available is used as an input for the fit, because in surface crystallography only a smaller number of structure factors can be obtained from the experiment, as compared to a bulk crystal structure analysis for a comparable number of atoms in the unit cell.

3.2.2 X-ray Reflectivity and Grazing Incidence Diffraction

A: X-ray Reflectivity
X-ray reflectivity is a standard technique to characterize layered structures with sub-nm resolution. In specular reflectivity experiments, the reflected intensity is recorded as a function of the incident angle μ with respect to the surface in symmetric condition ($\mu = \gamma$, $\delta = 0$, see Fig. 3.1c). For small angles (below 1° for most materials at a photon energy of 10 keV) total external reflection occurs and the Fresnel equations can be applied to calculate the reflected intensity [16–18]. In off-specular reflectivity experiments ($\mu \neq \gamma$, $\delta = 0$) an additional momentum transfer component is applied parallel to the surface along the incident beam direction. The combination of both techniques gives morphological information (independent of the crystallinity of the sample) with 0.1 nm resolution on: layer thickness, total-layer electron density, interfacial mean square roughness perpendicular to the surface, and lateral interfacial correlations [19] (height-height correlation function).
B: Grazing Incidence X-ray Diffraction
For small μ and/or γ (in the order of the critical angle α_c for total external reflection) and finite in-plane momentum transfer $Q_{||}$ (see Fig. 3.1b) the incident and the exit beam undergo strong refraction effects and the diffracted intensity can no longer be treated in kinematical approximation. The distorted-wave Born approximation has to be used instead, which includes refraction effects of the incident and the exit beam fully dynamically and the diffraction process itself in kinematic approximation. The following general expression can be derived for the scattered intensity [2]:

$$I(Q') \sim |T(\mu)|^2 \cdot |T(\gamma)|^2 \cdot S(Q') \,. \tag{3.2}$$

Here $|T(\mu, \gamma)|$ denotes the optical transmission function of the incident and exit beam, respectively [20]. Note that (3.2) is symmetric in μ and γ because of the reciprocity of the X-ray light path. $S(Q')$ denotes the kinematical structure factor which depends on the momentum transfer Q' inside the material. $S(Q')$ can describe any diffraction process, like Bragg scattering from near-surface lattice planes, small angle scattering, or interstitial diffuse scattering. Most important for grazing-incidence diffraction, the diffraction signal $S(\boldsymbol{Q}')$ can be obtained with a depth resolution from nm to μm, depending on the choice of the incident and the exit angles μ and γ [2]. Over the scattering depth, the component of the electrical wave field perpendicular to the surface is exponentially damped. For SXRD and SAXS the grazing-incidence geometry results in an improved signal-to-noise ratio.

C: Grazing Incidence Small Angle Scattering

For grazing incidence small angle X-ray scattering (GISAXS), an additional (small) momentum transfer $\boldsymbol{Q}_{\|}$ ($\delta \neq 0$) is present. For GISAXS, $\boldsymbol{Q}_{\|}$ is much smaller than typical reciprocal-lattice vectors from atomic planes. From GISAXS experiments combined information on nanoparticle size and distance distributions can be obtained [9]. In general, a quantitative analysis of GISAXS data is possible, but it is hampered by the sometimes difficult deconvolution of the distance and the size information of the nanostructures.

3.2.3 *In situ and* **Operando** *Sample Environments*

In this section the design of different in situ X-ray diffraction chambers is discussed, which allow on the one hand preparation of surfaces and nanostructures under controlled ultrahigh vacuum (UHV) conditions, and which are on the other hand compatible with atmospheric pressures thus bridging the pressure gap of 12 orders of magnitude from 10^{-9} mbar to 1 bar [21]. Two concepts are followed: For the first the chamber is operated at pressures above 10^{-3} mbar as batch reactor [22]. The second type of chamber can be run as a true flow reactor from about 10^{-2} mbar (when working with diluted gases) up to 1 bar [23].

Figure 3.2a shows a photograph of experimental equipment, available at beamline ID03 at the ESRF, that enables studies of model catalysts under higher pressures of a gas or semi-realistic reaction conditions in a batch reactor using SXRD [22]. The batch reactor consists of a 360° 2-mm-thick beryllium window with a geometry that allows operation of the chamber under external pressure of the atmosphere when the chamber is in vacuum and under internal pressure when the chamber is pressurized. The beryllium window is almost transparent to hard X-rays due to its low atomic number. The sample surfaces can be prepared in UHV with traditional surface-science recipes such as ion sputtering, annealing, etc. The batch reactor can be operated in a pressure range between 10^{-9} mbar and 5 bar to study surface structures of the adsorbed gases and possible modifications of the substrates induced by the surrounding gas. The sample can be heated by a ceramic heating plate and can reach a temperature of 1000 °C under vacuum and ~550 °C under 1 bar of pressure. The gas composition within the pressurized part of the chamber can be analyzed by a mass spectrometer simultaneously as the sample surface is probed by the X-rays. The chamber itself and its technical accessories are mounted onto a diffractometer to perform diffraction experiments with large scattering angles, allowing for exploring large regions of reciprocal space. Figure 3.2b shows the batch chamber in its high pressure oxidation variant. The sample sits inside a free standing, X-ray transpar-

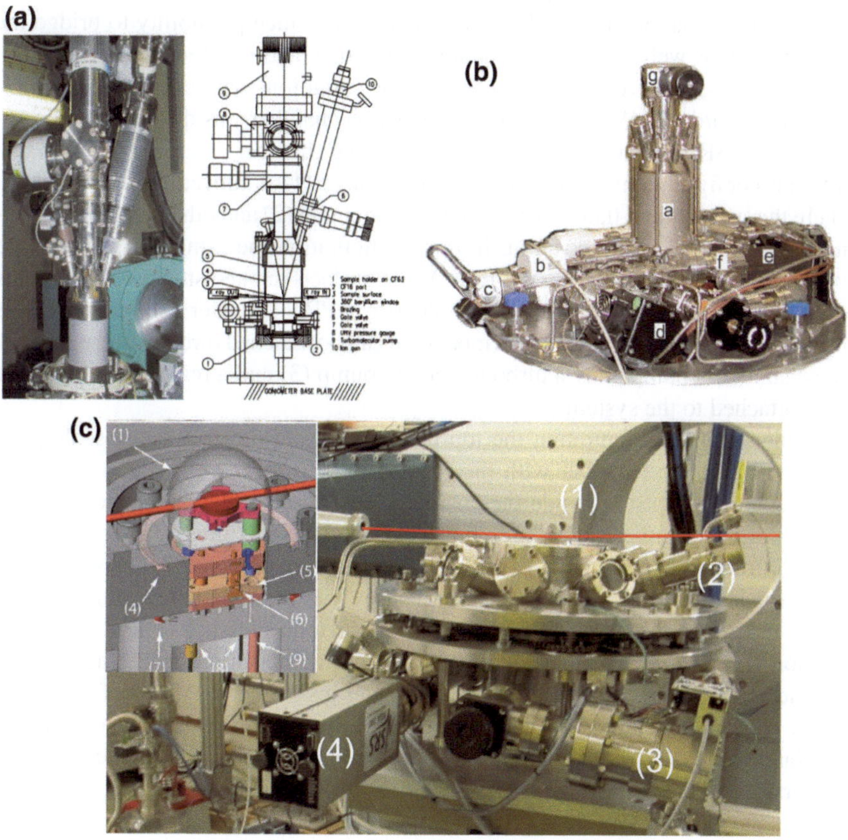

Fig. 3.2 **a** UHV-to-atmospheric-pressure X-ray diffraction chamber for batch catalysis experiments (adopted from [22]) **b** modified version of batch chamber for pure oxidation experiments **c** UHV-to-atmospheric-pressure flow reaction chamber mounted on the heavy diffractometer at P09, PETRA III (DESY). *Inset* schematic zoom of the reaction volume

ent 2-mm-thick Beryllium cylinder (a). The pressure is monitored from UHV up to 10^{-3} mbar by a cold cathode gauge (c) and up to 1 bar by two independent capacitive pressure gauges (b). The chamber is pumped via a turbo molecular pump (d) which can be isolated from the chamber by an angular valve. For controlled sample transfer from a preparation chamber to the synchrotron beamline the chamber can be pumped by an ion getter pump (e). Gases can be dosed through rough- and fine-adjustable leak valves. On top of the chamber, an ion sputter gun for surface cleaning or an evaporator can be mounted. The Al_2O_3-encapsulated Pt-wire heater allows sample temperatures up to ~900 K. The samples are mounted onto inconel sample holders and they can be brought into direct contact with a thermocouple for temperature control. A larger batch reactor (volume 5.5 l) for surface X-ray diffraction is discussed in [24].

For pure oxidation studies the batch chamber is a nice possibility to bridge the pressure gap, as well as for relatively slow catalytic reactions. Technologically relevant processes, however, take place under flow conditions with controlled gas mixtures. Therefore a UHV-compatible flow reaction chamber was developed at beamline ID03, ESRF, as pictured in Fig. 3.2c [23]. The chamber can be operated in two different configurations: In the first, the top flange (1) is moved up to a position, where the surrounding flanges carrying the sputter gun (2) and other auxiliary equipment look at the sample position. In this configuration the sample surface can be prepared in a routine way known from UHV surface-science studies. Also in this setup, the sample can be heated up to 1200 K by a PBN-C heater with direct sample mounting to the heater plate. A thermocouple can be attached directly to the sample. The chamber is pumped by a turbo molecular pump (3) and a residual gas analyzer (4) is attached to the system.

In the second configuration the top flange is moved down, thereby sealing off the top part of the chamber with the Be dome (5) from the UHV part below by a gasket-carrying piston inside the chamber. Two capillaries are connected to the top part of the chamber, allowing to flow controlled gas mixtures, while performing X-ray diffraction experiments. For the gas flow a computer controlled gas system is used, which allows gas mixtures of several gases with individual flow control (up to a total integrated flow of 200 ml/min), as well as total pressure control inside the reaction volume from 1 mbar to 1.3 bar. The gas composition inside the reaction volume can be detected by the mass spectrometer inside the UHV part of the chamber by controlled leaking in from the reaction volume. Similar setups now exist at the Nanolab at DESY, at the division of Synchrotron Radiation Research at Lund University and beamline SIXS at SOLEIL.

3.3 In situ Near-Atmospheric-Pressure Oxidation of Transition Metal Surfaces and Nanoparticles

In catalytic oxidation reactions of 3d, 4d, and 5d catalyst materials such as Cu, Ag, Pd, Rh, Ru, Pt, or Ir, it is frequently debated whether the oxide of the metal is the most active phase [25, 26]. These noble metals are difficult to oxidize at low, near UHV oxygen pressures, which explains why their interaction with oxygen has not been well studied in the past by traditional surface science techniques. As a prerequisite for reaction studies it is therefore important to investigate the oxidation behavior of noble metals at (near-)atmospheric pressures. Recent results discussed here focused on model systems such as low-index single-crystal surfaces. Increasing complexity was introduced by the investigation of vicinal surfaces and epitaxial nanoparticles with well-defined shape on single crystal oxide supports.

3.3.1 *Transition-Metal Low-Index Surfaces*

3.3.1.1 Oxidation of Low Index Ag

Silver is an important catalyst for several reactions, like ethylene epoxidation and the partial oxidation of methanol to formaldehyde [27]. Both reactions take place at atmospheric oxygen pressures and at temperatures from 500 to 900 K. The active oxygen species has been a matter of debate in the literature for a long time and it is still unclear up to date [28]. The interaction of oxygen with Ag(111) was studied by low-energy electron diffraction (LEED) already in the early seventies and a p(4x4) LEED pattern was observed [29–33]. Based on previous LEED, X-ray photoelecrton spectroscopy (XPS), and scanning tunneling microscopy (STM) investigations in combination with density functional theory (DFT) calculations the p(4x4) reconstruction was interpreted as an epitaxial $Ag_2O(111)$ layer, rotated by 30° with respect to the hexagonal unit cell of the Ag(111) substrate. The SXRD measurements together with new STM data disproved the $Ag_2O(111)$ layer model and gave evidence for a different structural model based on nanometric Ag(111) triangles, occupying faulted and unfaulted sites of the fcc stacking sequence [34, 35]. Oxygen ions occupy the furrows in between the Ag triangles in line with a chemisorption-induced reconstruction. The model is also supported by DFT calculations taking van der Waals interactions between Ag atoms into account. Furtheron the formation of epitaxial bulk oxide $Ag_2O(111)$ was observed after exposing the Ag(111) surface to atmospheric oxygen pressures at 800 K and cooling to room temperature [36]. The bulk oxide grows in coexistence with an ultrathin surface oxide layer, which very likely exhibits a trilayer surface oxide structure.

The (100) surface was exposed to near-atmospheric oxygen pressures at 440 K and it was studied by in situ SXRD [37]. A disordered form of a $p(2\sqrt{2} \times \sqrt{2})$ missing row reconstruction could be observed in the in situ SXRD experiment at 10 mbar O_2 pressure and 440 K. The surface undergoes a reversible transformation to the missing-row-like structure by increasing and decreasing the oxygen pressure. Under conditions, at which the missing row structure is observed, no evidence for the formation of subsurface, interstitial oxygen was detected. An ordered $p(2\sqrt{2} \times \sqrt{2})$ missing-row reconstruction was reported previously from low temperature LEED and X-ray photoelectron diffraction experiments [38]. Similar oxygen-induced order - disorder transitions were observed for the Cu(100) surface, as followed by SXRD [39].

3.3.1.2 Low-Index Pd Surfaces

Pd exhibits a high reactivity for CO oxidation, which is one of the basic catalytic reactions. As a prerequisite to CO oxidation under realistic reaction conditions, the interaction of oxygen with Pd surfaces up to atmospheric pressures needs to be understood. For Pd(100) and Pd(111) the formation of O-Pd-O trilayers is reported in the literature [40, 41]. Both have in common, that twofold and fourfold coordinated Pd atoms coexist (note, that in bulk PdO the Pd atoms are fourfold coordinated). On Pd(100) a surface oxide in a $(\sqrt{5} \times \sqrt{5})$ arrangement is formed, which exhibits a structure close to the (101) plane of PdO bulk oxide. One of the key questions of today's research in catalysis is whether chemisorbed oxygen or such surface oxide layers play a role for catalytic processes or nanometric layers of bulk oxide PdO instead [42–44]. Therefore, the stability of the surface oxide layers was studied as a function of the oxygen chemical potential μ [45–48]. Figure 3.3a shows the transition of the surface oxide to the bulk oxide for Pd(111) as a function of the oxygen pressure at 673 K. The surface oxide forms a $\sqrt{6} \times \sqrt{6}$-like overlayer with Pd_5O_4 stoichiometry on Pd(111) [40], which gives rise to a distinct surface X-ray diffraction signal. The signal from the epitaxial PdO bulk oxide can be discriminated as well. The results from such pressure- and temperature-dependent measurements are summarized in the stability diagram in Fig. 3.3b. The transition from the surface covered with chemisorbed oxygen to the surface oxide can be described by a line of constant oxygen chemical potential [48] of –1.24 eV, indicating local equilibrium. On the contrary, the bulk oxide formation is kinetically hindered over a wide temperature range and it does not take place below 500 K because the surface oxide is

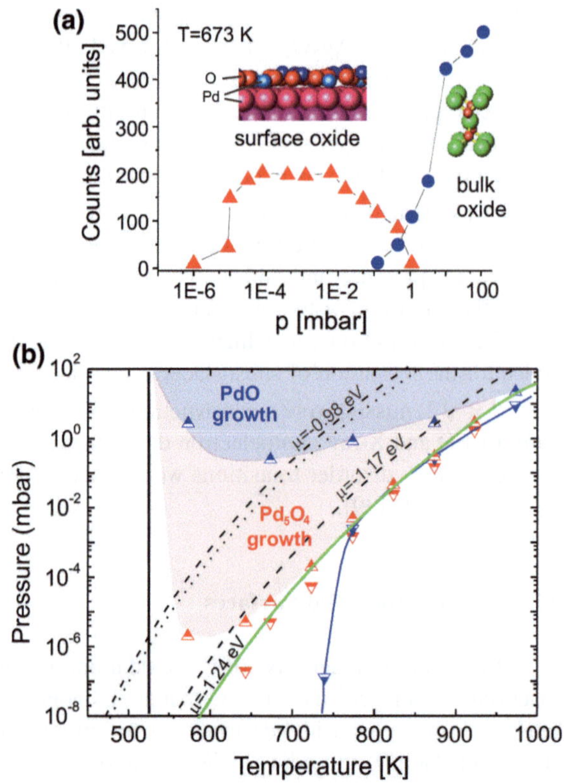

Fig. 3.3 **a** Experimental observation of the surface oxide to bulk oxide transition at constant temperature as a function of the oxygen pressure (*triangles*: surface oxide signal, *circles*: bulk oxide signal), **b** experimental stability diagram derived from pressure dependent measurements at different temperatures (*upward triangles*: oxygen pressure increasing, *downward pointing triangles*: oxygen pressure decreasing). From [48]

passivating the surface. At 973 K kinetic barriers are lifted and the bulk oxide forms directly without the intermediate surface oxide. Similar in situ SXRD experiments were performed on the Pd(110) surface, pointing towards a reduction of the kinetic barriers for bulk oxide formation, as compared to Pd(100) and Pd(111) [49].

3.3.1.3 Oxidation of Low Index Rh Surfaces

Rh oxides involved in catalytic oxidation reactions typically form at oxygen pressures of 10^{-3} mbar and above at application-relevant temperatures. The Rh(100) surface undergoes a transformation from a p(2×2) to a p(3×1) reconstruction with increasing oxygen exposure. Close to the chemical potential for bulk oxide formation, a c(2×8) reconstruction is observed [50]. A detailed atomistic model of this structure was determined by a combination of DFT calculations, surface X-ray diffraction, LEED, high-resolution core-level spectroscopy (HRCLS), and STM [51]. The structural model is based on a hexagonal oxygen-Rh-oxygen trilayer, similar to the trilayers formed on Pd(100) and Pd(111). The hexagonal layer is slightly distorted to accommodate a c(2×8) structure and it forms in two domains rotated by 90°. Figure 3.4a shows the the structural model of the trilayer surface oxide which

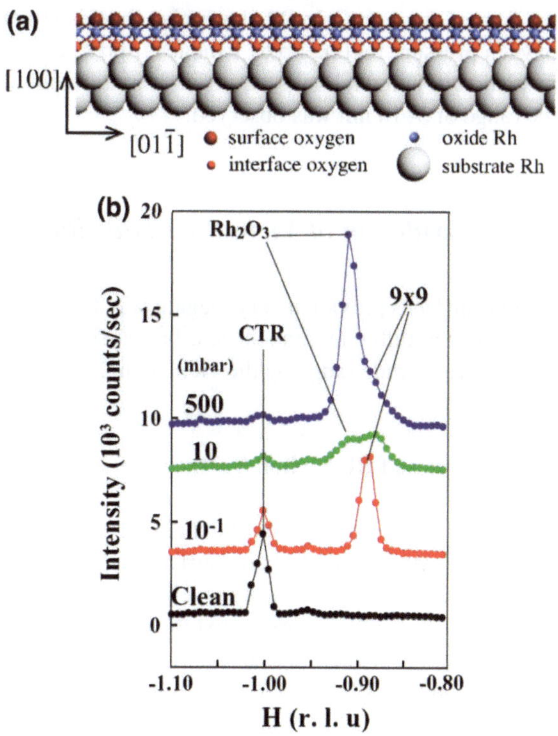

Fig. 3.4 **a** Structural model for the surface oxide layer on the Rh (100) surface, **b** SXRD data showing the transition from the Rh surface oxide to Rh_2O_3 bulk oxide on Rh(111). Adopted from [45]

is also observed on Rh(110) and Rh(111), pointing towards the high stability of this structure [45, 52]. The transition of the surface oxide on Rh(111) to Rh_2O_3 bulk oxide as a function of the oxygen pressure at 800 K is illustrated in Fig. 3.4b: In scans with momentum transfer in the surface plane the surface oxide signal gives rise to a reflection at H = 0.89 reciprocal lattice units at 10^{-1} mbar, which shifts to H = 0.91 for the Rh_2O_3 bulk oxide at 10 mbar O_2 pressure [45].

3.3.1.4 Oxidation of Pt(111) and Ir(111)

Pt and Ir are important 5d transition-metal catalysts for CO oxidation [53, 54]. In addition, Pt coatings are discussed as protective layers in aggressive chemical environments, for example to protect the electrodes of sparkling plugs in cars and to increase their lifetime. The oxidation of a Pt(111) single crystal and epitaxial, (111)-oriented Pt films on α-Al_2O_3 were studied by in situ surface X-ray diffraction at near-atmospheric pressures [55]. The oxidation of the Pt(111) surface is kinetically strongly hindered and an ultrathin α-PtO_2 layer forms at 910 K at 0.5 bar O_2 pressure. A structural analysis reveals that the oxide consists of one α-PtO_2 unit cell, which is compressed in the direction perpendicular to the surface and distorted in the surface plane.

A detailed SXRD study of the oxidation of the Ir(111) surface gave evidence of the formation of an O-Ir-O trilayer also for this system at 575 K in an intermediate oxygen pressure regime up to 1 mbar O_2 pressure [56]. At higher temperatures the formation of epitaxial bulk IrO_2 islands in different orientations with thicknesses in the range of 5–10 nm was observed.

3.3.2 *Oxidation of Vicinal 4d Transition-Metal Surfaces*

Vicinal surfaces provide a very elegant way to mimic defects such as steps and corners between different nanoparticle facets, since the geometry of the defect can be chosen over a wide range by the vicinal angle and orientation.

3.3.2.1 Pd and Rh Vicinal Surfaces

The oxidation of Pd(553) and Pd(112) vicinal surfaces was studied employing the powerful combination of in situ surface X-ray diffraction, high resolution scanning tunneling microscopy, and core level spectroscopy, together with DFT calculations [57, 58]. The fcc(553) surface exhibits threefold coordination on terrace and step sites whereas the (112) surface shows (100) type steps with fourfold symmetry. Both the (553) and (112) Pd surfaces are found to be stable after preparation under UHV conditions. At 600–700 K and 10^{-6} mbar oxygen pressure the (553) surface transforms into (332) facets, which are decorated by a surface oxide layer, exhibiting a

12 times larger unit cell along the terraces as compared to the substrate resembling the $\sqrt{5}$ trilayer forming on the Pd(100) surface. At pressures of 1 mbar epitaxial and polycrystalline bulk oxide formation sets in. The epitaxial oxide islands are tilted with respect to the substrate and grow in PdO(012) orientation. The morphology of the Pd(112) surface is also strongly influenced by the oxidation conditions: At 673 K, upon exposure to oxygen at pressures from 2×10^{-8} to 5×10^{-5} mbar, the (112) surface undergoes a massive rearrangement and (113)- and (335)-type facets are formed. Further increase of the O_2 partial pressure leads to a new rearrangement into (111)- and (113)-type facets. Bulk oxide formation is observed at 5 mbar O_2 pressure and T = 523 K, which is below the bulk oxide formation temperature of low-index (100) and (111) surfaces. This elucidates the role of steps towards lifting of kinetic barriers for bulk oxide formation.

For comparison, the oxidation of the Rh(553) surface was investigated by SXRD from UHV to near-atmospheric pressures [59]. At 380 °C and 10^{-6} mbar O_2 pressure the formation of (331) facets is observed along with an oxygen-induced restructuring along the steps. At higher pressures above 10^{-3} mbar and a temperature of 500 °C the formation of the O-Rh-O trilayer surface oxide stabilizes larger (111) facets, in line with the particular stability of this interface.

3.3.3 *Nanoparticle Oxidation Across the Pressure Gap: Pd, Rh and Pt Nanoparticles on MgO(100) and $MgAl_2O_4$(100)*

State of the art in situ structural studies of catalytic reactions are performed on powder samples made up by sub-micron grains of supporting oxide particles covered with nanometer-sized metal/alloy particles [60]. To obtain a deeper insight into the basic processes occurring during chemical reactions on nanoparticles, one possible approach is to replace the real catalyst by epitaxial nanoparticles with a well-defined orientation relationship and size distribution, grown by physical vapor deposition onto single crystal oxide supports [61]. Such a system allows *operando* X-ray studies under realistic pressures at elevated temperatures, giving detailed insight into structural and morphological changes during the reaction. Fcc metals like Pd, Rh and Pt grow on MgO(100) and $MgAl_2O_4$(100) in the (100) direction with truncated octahedral shape exposing mainly (111)- and (100)-type facets under UHV conditions.

Using reciprocal space mapping, Pd nanoparticles on MgO(100) with diameters in the range of 5–9 nm were shown to form nanometer sized (112)-type facets under 10^{-5} mbar oxygen exposure and 570 K, which is reversible under CO exposure [62]. In contrast, for Rh nanoparticles (100)- and (111)-type facets are stabilized under similar conditions by the formation of a surface oxide trilayer O-Rh-O shell, which was detected in corresponding X-ray diffraction line scans [63]. The Rh surface oxide formation is accompanied by a nanoparticle shape change increasing the (100) sur-

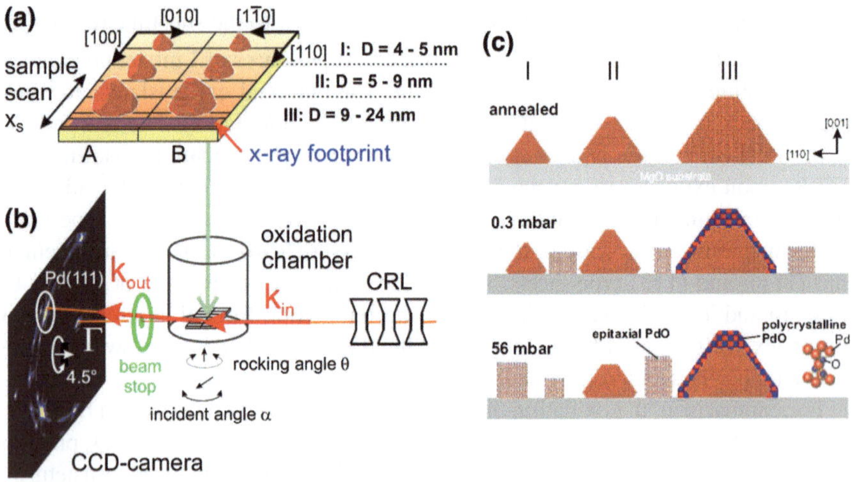

Fig. 3.5 Oxidation of Pd nanoparticles [65]. **a** Combinatorial sample architecture with nanoparticle stripes of different average diameter. **b** High energy X-ray diffraction scheme using a 2D detector. **c** Oxidation scenario for different nanoparticle size regimes

face area, which can be rationalized on the basis of the higher stability of the surface oxide on the Rh(100) surface [64]. Also in this case, the shape change is found to be reversible under CO exposure.

For the systematic investigation of size- and composition-dependent phenomena, a novel approach was developed combining high-energy grazing incidence X-ray diffraction with a combinatorial sample architecture. The size-dependent oxidation of epitaxial Pd nanoparticles on MgO(100) was investigated in the regime from 4–24 nm, revealing that the formation of a polycrystalline passivation layer is only observed for particle sizes above 9 nm [65]. For smaller particles the growth of epitaxial PdO is found instead (Fig. 3.5).

Alloying of Rh with Pd leads to a very different oxidation behavior: The composition-dependent oxidation of Pd-Rh nanoparticles on $MgAl_2O_4$(100) in the size regime 5–10 nm gave evidence for a preferential Rh oxide formation accompanied by oxidation-induced Rh surface segregation [66]. Pd oxide formation was only observed for pure Pd nanoparticles. For epitaxial Pt nanoparticles on MgO(100) the formation of higher-index facets was observed after oxidation at 570 K in the pressure regime of 10^{-3} mbar to 0.5 bar, which was found to be non-reversible under CO exposure in contrast to the Pd and Rh case [67]. The formation of epitaxial PtO_2 and Pt_3O_4 was observed, which was stable under vacuum annealing up to 923 K, pointing to strong kinetic barriers for their reduction.

3.4 In situ Catalytic Studies Using Batch Reactors

In this chapter examples of in situ catalytic reactions using a batch reactor by combining SXRD and mass spectrometry will be described. Here we will focus on Pt, Ru, Pd, and Rh surfaces during CO oxidation reactions as well as the Ru surface during HCl oxidation. As pioneering work it was demonstrated that the CO induced (2x1) reconstruction on Ni(110) is at room temperature stable up to atmospheric pressures [68]. The hydrogenation of butadiene over a $Ni_{0.92}Pd_{0.08}(110)$ model catalyst was investigated in batch mode [69], but this is not discussed in detail here. At the end of the section, results on batch experiments of model nanoparticle catalysts on single crystal oxide supports are presented.

3.4.1 CO Oxidation Over Pt

The first in situ SXRD CO-oxidation experiment using a batch reactor and a single crystal surface as a model catalyst was presented in 2005 [70]. In this study, a Pt(110) surface during CO oxidation at pressures up to 0.5 bar and temperatures up to 350 °C was studied. Prior to the CO oxidation experiments, the initial oxidation of the surface was investigated. At an oxygen pressure of 500 mbar and a sample

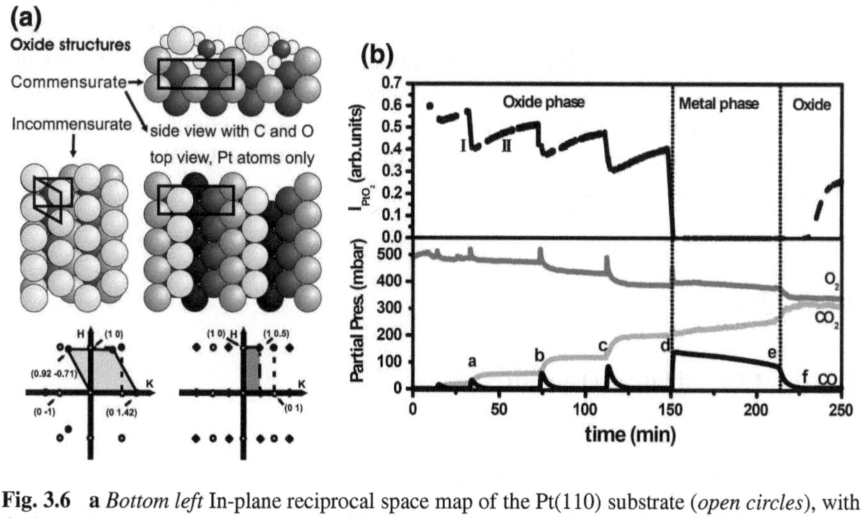

Fig. 3.6 **a** *Bottom left* In-plane reciprocal space map of the Pt(110) substrate (*open circles*), with the incommensurate, quasihexagonal oxide overlayer (only Pt atoms are shown, *solid circles*). *Bottom right* The in-plane reciprocal space map of the (1×2), commensurate oxide overlayer (*crosses*). The ball models show the real-space structures of the two oxides. **b** Simultaneously measured X-ray diffraction intensity at (0, 1.42, 0.5) from the quasihexagonal oxide (*top panel*) and partial pressures of CO, O_2, and CO_2 (*bottom panel*). Separate CO pulses were admitted to the reactor, which was initially filled with 500 mbar of O_2 at a temperature of 350 °C. The sharp peaks in P_{O2} are an artefact due to the sudden increase of total pressure at each CO pulse. From [70]

temperature of 350 °C, the formation of an incommensurate quasihexagonal α-PtO_2 oxide layer was observed. The structure found is described in Fig. 3.6. When CO was introduced together with oxygen, the catalytic oxidation of the CO into CO_2 could be observed. The experiment is shown in Fig. 3.6b, in which the intensity of the oxide peak (top part) is probed simultaneously with the O, CO and CO_2 pressures as a function of time. When the oxidized Pt surface is exposed to relatively moderate amounts of CO (Fig. 3.6b, "a" to "c"), the oxide layer is roughened by the reaction, but the surface remains oxidized and exhibits a high reaction rate. When exposed to higher pressures of CO (Fig. 3.6b "d"), the oxide is completely reduced, as indicated by the section termed metallic phase in Fig. 3.6b. A simultaneous decrease in the reactivity was observed. After point "d", the CO is consumed in the batch reactor by the oxidation into CO_2, and the CO/O pressure ratio is continuously reduced, since two CO molecules are consumed by one O_2 molecule. When the CO pressure is sufficiently low, the surface re-oxidizes, which can be seen by the re-appearance of the diffraction signal from the incommensurate α-PtO_2. However, the reaction rate calculated from the increase in the CO_2 signal increases sharply about 20 min before the re-appearance of the oxide signal. Instead, a set of new diffraction peaks could be detected simultaneously with the increase in reaction rate (not shown). These peaks were found to correspond to a commensurate (1×2) surface structure. The structure could only be observed under reaction conditions, any attempt to quench the structure by cooling and evacuating the chamber resulted in the disappearance of the diffraction spots. By structural measurements during the reaction combined with DFT calculations, the commensurate structure shown in Fig. 3.6a (right) could be proposed, which involves the presence of one carbonate ion per unit cell. These observations show that in situ measurements under actual reaction conditions are crucial for a meaningful investigation of the surface structure and chemical behavior of this model catalyst. Clearly, the presence and role of the commensurate (1×2)-layer could not have been found either in experiments under UHV or in so-called pre- and post-reaction experiments.

3.4.2 Ru

3.4.2.1 CO Oxidation Over Ru

The oxidation and reduction of a Ru(0001) surface was investigated by He et al. [71] paving the way for CO oxidation experiments over the Ru(0001) surface at realistic conditions using SXRD. The catalytic CO oxidation over the Ru(0001) surface was investigated in detail using the above batch reactor [72, 73]. In essence, the studies were conducted to compare the CO_2 production in the presence of the so-called metallic phase and in the presence of an oxidized Ru(0001) surface, the $RuO_2(110)$ surface.

In a first set of experiments the investigations were made under isothermal conditions [71]. An example of the experiments is shown in Fig. 3.7. The temperature was

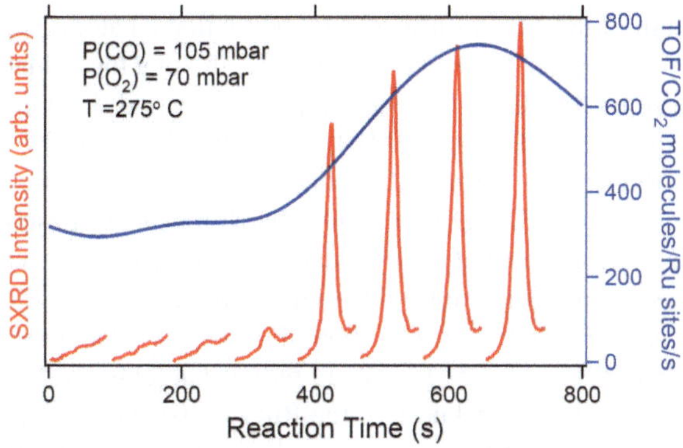

Fig. 3.7 The activity of CO oxidation and the surface structure of the Ru(0001) surface during isothermal conditions. The experiment starts with $P(O_2) = 70$ mbar and $P(CO) = 105$ mbar at a sample temperature of 275 °C. From the CO_2 partial pressure and a gas temperature of 25 °C the TOF can be determined. The $RuO_2(110)$ can be identified after about 200 s in the repetitive l-scans at h = 0.73. The appearance of the $RuO_2(110)$ reflection can be correlated to an increase in the TOF. From [74]

kept constant at 275 °C with initial O_2 and CO partial gas pressures of 70 and 105 mbar, respectively. At the beginning of the experiment, the activity is low, but not zero. Since twice as many CO molecules are consumed compared to O_2, the environment becomes increasingly oxidizing, and after 200 s the surface transforms from the metallic state into the $RuO_2(110)$ state, in conjunction with a significant increase of the CO_2 production and turnover frequency (TOF). This observation indicates that the presence of a $RuO_2(110)$ phase results in a more active phase than the non-oxidized surface.

Similar experiments were done at different temperatures, and it was found that the metallic and the $RuO_2(110)$ phase had an approximately equal TOF at 255 °C but that the $RuO_2(110)$ phase had a higher TOF at 275 °C and higher temperatures. Furthermore, a significant reaction-induced temperature increase of the sample was observed at temperatures of 275 °C and higher. In a second set of experiments [72] the effect of changing the temperature of the sample was investigated. Also in this study it was concluded that the metallic and the $RuO_2(110)$ phases were almost equal at temperatures below 245 °C while at higher temperatures the $RuO_2(110)$ was more active.

3.4.2.2 HCl Oxidation Over Ru

The oxidation of hydrochloric acid (HCl) by oxygen (O_2) into water (H_2O) and chlorine (Cl_2), the so-called Deacon process [75], is an important reaction for the industrial chlorine production. Recently, it was observed that RuO_2-covered TiO_2 leads to an efficient catalyst for the Deacon process [76]. Fundamental studies of the HCl interaction with the $RuO_2(110)$ surface have shown that Cl atoms replace the bridging oxygens in the top-most $RuO_2(110)$ layer [77], and that the oxidation of HCl into H_2O and Cl_2 appears to follow the Langmuir-Hinshelwood (LH) mechanism along the rows of under-coordinated Ru sites [74, 78], at least under UHV conditions. The stability of the $RuO_2(110)$ during HCl oxidation at more realistic conditions was studied [79] using in situ SXRD in a similar batch reactor as described above, the difference being a replacement of the cylindrical X-ray Be window with a 1 mm thick Al window, to avoid to brittle the Be due to high HCl exposures. It could be shown that in pure HCl at a pressure of 1 mbar both the $RuO_2(110)$ and the $RuO_2(100)$ are stable up to a temperature of 325 °C, but are reduced at temperatures above 325 °C. By introducing oxygen, the thickness of the $RuO_2(110)$ and $RuO_2(100)$ films were even observed to increase at temperatures as high as 345 °C at oxidizing reaction conditions. By combining the SXRD data with online mass spectrometry the mean turnover frequencies could be determined in the presence of the $RuO_2(110)$ and $RuO_2(100)$ surfaces, indicating that the HCl-oxidation is structure insensitive.

3.4.3 CO Oxidation Over Rh

The CO oxidation over Rh catalysts has been studied for a long time under UHV [80, 81] as well as under more realistic conditions [82]. However, very little is reported on the surface structures of Rh model catalyst surfaces under more realistic pressure conditions.

The surface structure over Rh(111) and Rh(100) surfaces during CO oxidation at elevated pressures and temperatures was studied by SXRD combined with mass spectrometry using the batch reactor as described above [83, 84].

Previous oxidation experiments have shown that the same surface oxide forms on Rh(111) [45], Rh(100) [51], and Rh(110) [52] surfaces as well as on vicinal Rh surfaces [59, 85]. In addition, the same surface oxide has been found to form on PtRh(100) [86] as well as on Rh nanoparticles [63, 87, 88]. The structure of the trilayer surface oxide is shown in Fig. 3.8a.

Armed with the information from the oxidation studies, the expected reciprocal space for the surface oxide on Rh(111) is shown in Fig. 3.8b. Using a point detector, the scan shown in Fig. 3.8c (left) across the surface oxide allows to detect the presence of the surface oxide combined with online mass spectrometry at a constant temperature of ~240 °C. The results are shown in Fig. 3.8d–g. In Fig. 3.8d the partial pressures of O_2, CO, and CO_2 are shown. Initially, a partial O_2 pressure of 300 mbar was introduced in the reactor, and at time t = 0 s 300 mbar of CO was added to the reactor, starting the CO oxidation and the formation of CO_2. The CO_2 partial pressure is increasing linearly at the same time as the CO and O_2 are decreasing as expected. After around 300 s a sharp increase in the CO_2 partial pressure is observed in conjunction with sharp decreases of the O_2 and CO pressures. The CO_2 production shown in Fig. 3.8e displays only a weak increase until t = 3300 s, when there is a sudden strong increase of the CO_2 production. In Fig. 3.8f, the surface oxide

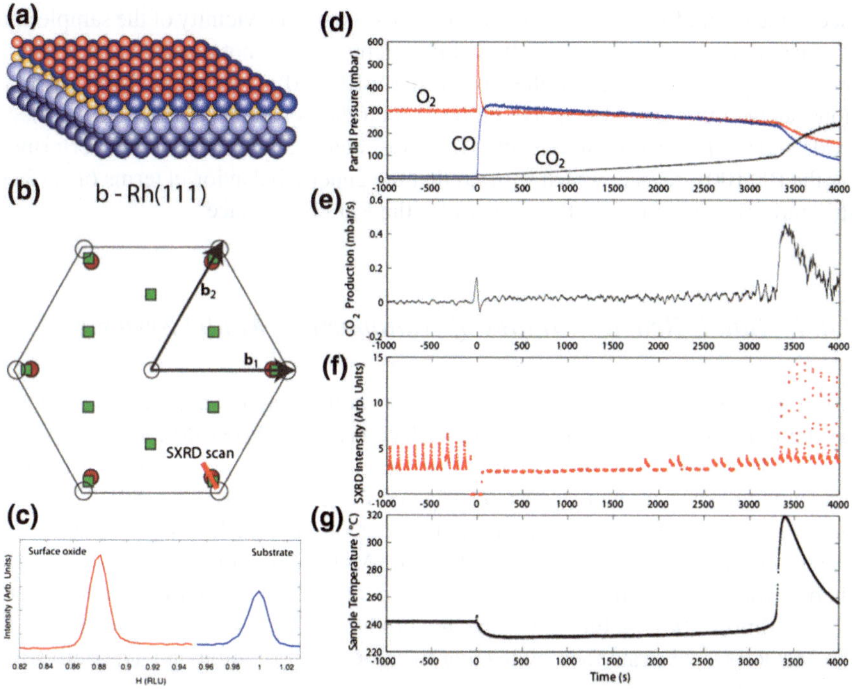

Fig. 3.8 **a** Model of the surface oxide found on all investigated Rh surface orientations, here shown on top of a (111) surface. **b** Reciprocal space map for Rh(111), including the surface oxide as well as the corundum-structured bulk oxide. **c** Line scan along the *red line* in (**b**). **d** partial pressures of O_2, CO, and CO_2. **e** CO_2 production as derived from (**d**). **f** consecutive SXRD scans along the *red line* in (**b**), showing the presence or absence of the surface oxide. **g** sample temperature during the experiment. From [83]

scans, as shown in Fig. 3.8c (left), are shown. When only oxygen is present at this temperature, the surface oxide can be observed, but as soon as the CO is introduced at t = 0 s, the diffraction from the surface oxide disappears. At around t = 3300 s, the surface oxide peak re-appears, in conjunction with the sharp increase in CO_2 production as observed using mass spectrometry. Finally, the sample temperature is shown in Fig. 3.8g, showing that the temperature is decreasing due to the introduction of the CO at t = 0 s, a slow increase until t = 3300 s is observed, at which a strong increase of the sample temperature is detected, indicating heating due to the exothermic nature of the CO-oxidation reaction.

The slow increase of the CO_2 production from t = 0 s to t = 3300 s can be related to a gradual decrease of the CO poisoning of the surface as the gas composition is slowly changed in the reactor. At t = 3300 s, CO suddenly desorbs from the surface and the reaction rapidly becomes mass-transfer-limited (MTL) instead of reaction-limited. In fact, at this point almost all CO in the vicinity of the sample is converted into CO_2 resulting in the sharp peak as observed in the mass spectrometry. The surface oxide is rapidly formed due to the excess oxygen in the vicinity of the sample and the increased temperature. From these measurements, it is not possible to determine the active phase, however, in the case of the PtRh(100) experiments [86], calculations showed that the most active phase was on the border between the Rh surface oxide and the Pt surface underneath. An almost identical experiment was performed for the Rh(100) surface, which showed that the general behavior in terms of surface structure vs. reactivity was the same as for the Rh(111) surface.

3.4.4 Batch Reactor Studies of Nanoparticle Model Systems

A limited number of batch reaction studies on well-defined nanoparticles on single crystal oxide supports combined with surface sensitive X-ray diffraction was performed: In a systematic investigation the nanoparticle size dependence of the CO oxidation turnover frequency was investigated for Au on TiO_2(110) [89]. The highest reactivity was found for a particle diameter of 2.1 nm. In a batch reactor CO oxidation experiment over Pd nanoparticles on MgO(100) with truncated octahedral shape it was observed that carbon is dissolving into the Pd nanoparticles under reaction conditions [90]. At the same time, the produced CO_2 was found to interact with the MgO(100) substrate thereby leading to carbonate formation and a de-activation of the catalyst.

3.5 *Operando* Studies Under Flow Conditions

This section will give a summary of the current state of research of X-ray diffraction experiments using the flow setup described in Sect. 3.2.3. Again, almost all of these studies concern CO oxidation over Pd, Rh, and Pt, but also include a study of methane oxidation over Pd(100).

3.5.1 CO Oxidation Over Pd(100)

As discussed above, there is a debate going on concerning the active phase of Pd(100) during catalytic CO oxidation in O_2 excess. The first use of the flow reactor at ID03 was also directed towards this debate [91, 92].

In [91], a set of measurements is discussed under conditions that were as similar as possible to those used in [93] when a "hyper-active phase" between the CO-poisoned metallic state and the mass-transfer-limited regime was identified. The results were very closely reproduced, with the difference that there were no signs of any extra active phase in between the metallic and mass-transfer-limited regimes. In [93] it was concluded that the most active phase is metallic, covered by adsorbed oxygen, while the SXRD measurements in [91] clearly showed that the surface is oxidized when the catalytic activity is high.

In a follow-up study the phase diagram of Pd(100) under different reaction conditions was mapped out [92]. The clean sample was exposed to a flow of 50 ml_n/min (1 ml_n is the number of molecules corresponding to 1 ml at standard pressure and temperature) with different CO to O_2 ratios using Ar as a carrier gas at a total pressure of 200 mbar. Simultaneously the surface phase was monitored in situ by SXRD. Using a 2D detector, a common reflection of the PdO surface and bulk oxides was monitored.

The resulting phase diagram is shown in Fig. 3.9. The markers indicate more exactly under which conditions the measurements were done, and whether the surface is metallic (solid symbols) or oxidized (open symbols). To the right of the solid zig-zag line, the surface is metallic and the catalytic activity is relatively low, but increases with temperature. To the left of this line, the activity of the sample is high enough for the measurements to be mass-transfer-limited, and SXRD always reveals an oxide. The nature of this oxide, however, depends on the conditions. The conclusion drawn from these measurements is that, as soon as the activity of the sample is high enough to be mass-transfer-limited, the surface also exposes an oxide. Hence it was argued that the oxide phases must be more active than the metallic surface.

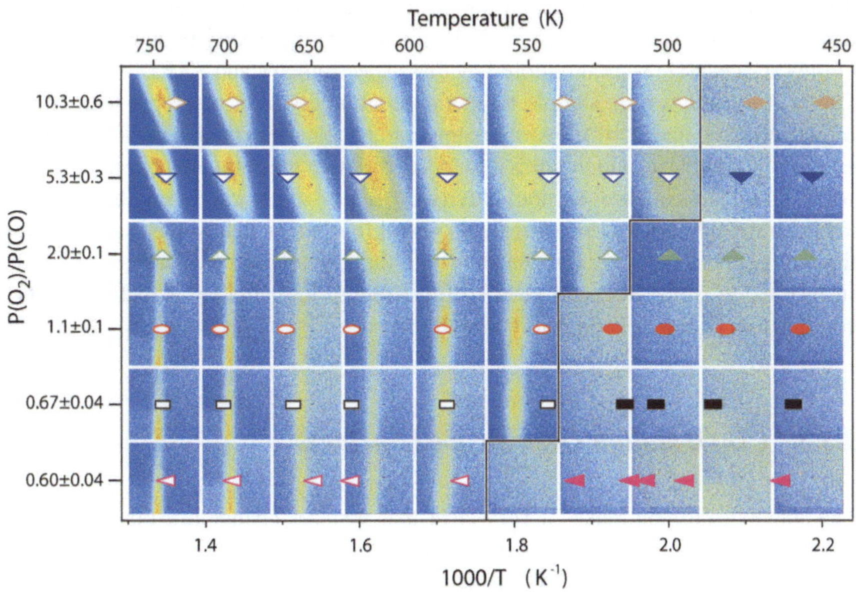

Fig. 3.9 Recorded diffraction signal as a function of the $P(O_2)$:$P(CO)$ ratio and sample temperature. The total gas pressure in the reactor was kept constant at 200 mbar, the partial CO pressure was kept constant at 10 mbar. Oxygen and argon were mixed in the flow to obtain the desired O_2:CO ratio. *Open symbols* denote the presence of an oxide, *filled symbols* denote that the surface was metallic in nature. The *thick black line* indicates the boundary between the metallic and oxidic phases. From [92]

Another interesting phenomenon in catalysis are self-sustained reaction oscillations, which were studied during steady-state CO-oxidation over Pd(100) [94]. At a sample temperature of 447 K and an incoming gas flow of 500 mbar O_2, 25 mbar CO, and 675 mbar Ar, the CO_2 production rate is found to oscillate spontaneously between the MTL and very low activity. As above, *operando* SXRD reveals that the surface is oxidized in the high-activity regime and metallic in the low-activity regime. In addition, the authors have analyzed the width of the diffraction peaks and correlate it to the roughness and number of steps at the surface. In the highly-active, oxidized phase, the surface is found to get rougher and rougher, until the activity switches. In the following low-activity metal phase, the surface smoothens again, until the next switch to high activity. The authors therefore conclude that the oscillations are controlled by the step density at the surface. A high number of steps stabilizes the low-activity metal surface, while a smooth surface more easily oxidizes into the high-activity phase.

3.5.2 CO Oxidation Over Rh(111)

Also CO oxidation over Rh(111) has been studied in the flow reactor [95, 96]. In flows of different gas mixtures, around a stoichiometric mixture of CO and O_2, the Rh(111) sample was heated and cooled while the catalytic activity and the presence of the surface oxide were monitored by mass spectrometry and SXRD, respectively. The activity follows the sample temperature smoothly until a sample temperature of between 200 and 250 °C. At this temperature, the experiments performed in an excess of O_2 show a sudden jump into a highly active mass-transfer-limited regime, while in a CO-rich gas flow, the activity continues to increase smoothly with the temperature. The SXRD measurements reveal that the switch to the mass-transfer-limited regime (in O_2 excess) coincides with the appearance of a surface oxide. The surface-oxide signal does not grow to full strength immediately, which suggests a combination of low coverage and small domains.

3.5.3 CO Oxidation Over Stepped Surfaces

The above discussion makes it very clear that surfaces change drastically with the exposure to reaction gas mixtures. This is not less true for stepped surfaces, which are used to investigate the effect of defects on catalytic reactions [97, 98]. Since CTRs are always perpendicular to the surface, the appearance of new facets is revealed by the appearance of CTRs in new directions. Figure 3.10 shows a map of reciprocal space around the Pt(977) Bragg reflections at $(H, K, L) = (16, 0, 5)$ and $(18, 0, 28)$ [97]. Vertically from these Bragg reflections there are two CTRs revealing the overall (553) orientation of the surface. In addition, there is one CTR connecting the Bragg reflections, and hence leaning slightly. This CTR is perpendicular to the (111) facets

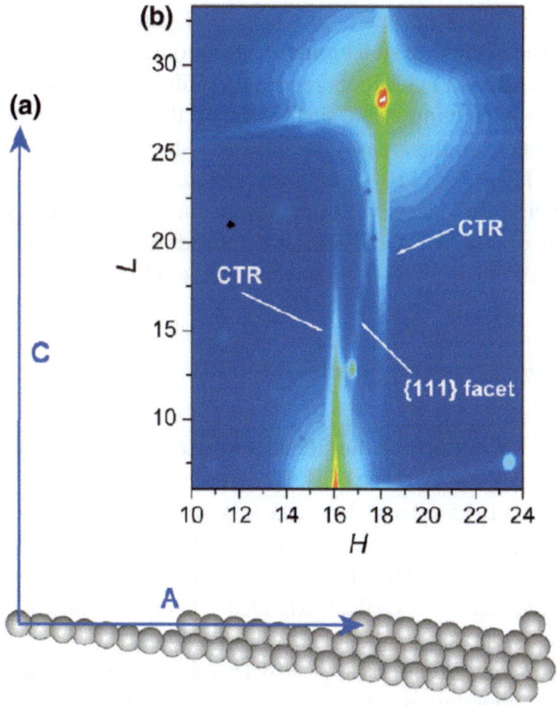

Fig. 3.10 **a** Side view of the (977) surface and representation of the A and C vectors in direct space. C is perpendicular to the 997 planes and A has a length twice the step separation. **b** Diffracted intensities in a portion of the plane (H,L) at K = 0 in reciprocal space. The H and L axes are parallel to vectors A and C respectively. The intense *red spot* at (H, L) = (18, 28) arises from a Bragg reflection from the bulk of the crystal. The diffuse intensity streak emanating from the Bragg reflection, parallel to the L axis, is a crystal truncation rod (CTR) of the 977 surface plane. At (H, L) = (16, 5) there is another Bragg reflection which tail is visible in the lower part of the figure. Again, its associated diffraction rod from the stepped surface parallel to the L axis is noticeable. In addition, an inclined diffuse intensity line connecting the two bulk Bragg reflections is visible. It corresponds to a diffracted rod from (111) surface planes indicative of the existence of surface (111)s facets. From [97]

that make up the terraces between the steps. For a perfect (977) surface, the (111) CTR should cancel out, but there are always small variations in the terrace size, such that for the clean surface a combination of (997) and (111) CTRs was found. During this study, the evolution of the (111) and (997) facets is followed by line scans revealing the intensity of the corresponding CTRs, while the reaction gas mixture is varied. In a stoichiometric mixture of CO and O_2 the (977) facet becomes more visible, while excess of either CO or O_2 causes rearrangements of the surface into larger (111) facets. In order for the macroscopic surface orientation to remain in the (977) direction, the growth of (111) facets must be accompanied by step bunching, but the structure of the areas where the steps bunch together is unclear from the present study.

Similar observations were made for the Rh(553) surface [98]. Analogous to the Pt(977) case above, in close to stoichiometric gas mixtures with slight CO excess, the presence of (553) and (111) CTRs is reported, while larger excess of CO or excess of O_2 both results in faceting of the surface. Here, the 2D maps make it possible to find out what happens in the areas with increased step intensities. In CO excess the formation of a CTR in the (110) direction is found, while O_2 excess results in new (331) rods, showing that facets with the corresponding orientations are formed. In addition, at large excess of O_2, there is an extra rod in the (111) direction, showing the formation of the surface oxide on the enlarged (111) terraces.

3.5.4 Methane Oxidation Over Pd(100)

Another example of the use of the 2D detector to identify the active phase of a Pd(100) model catalyst is methane (CH_4) oxidation [99]. In relatively low pressures ($P(CH_4)$:$P(O_2)$ = 0.1:0.5 mbar), the surface is always inactive in the oxide phase and the activity starts when the temperature is high enough in order to transform the surface into a metallic state. The 2D diffraction map of the oxide, both in the beginning and in the end, shows a vertical extended reflection, which is indicative of an ultrathin surface oxide.

At higher pressure ($P(CH_4)$:$P(O_2)$ = 6:16 mbar), however, there is an active regime at lower temperatures where the surface is still oxidized, but the activity drops before it increases again when the surface turns metallic. As the temperature increases, the detector images evolve from a vertical line with rather smooth intensity through a vertical line with intensity more and more concentrated in one spot, to a strongly leaning line with a very strong maximum. This reflects the development of the oxide from a surface oxide through an epitaxial bulk oxide into a polycrystalline oxide structure. Theoretical calculations have predicted that an oxide growing on Pd(100) will expose PdO(101)-oriented surface facets, while PdO on its own (or a thick enough film on Pd(100)) will expose (100) oriented facets. Hence the following conclusions can be drawn from the measurements: They show the transformation from a surface oxide to an epitaxial oxide exposing (101) facets, and finally the oxide gets thick enough to lose its registry with the substrate, becomes polycrystalline and exposes (100) facets. This would also explain the activity variations, as PdO(101) exposes undercoordinated Pd atoms that are predicted to be the active sites, while PdO(100) does not expose such sites.

3.5.5 The Use of Large 2D Detectors in Combination with High X-ray Photon Energies

Finally, surface-sensitive X-ray diffraction is right now going through a revolution with the use of large 2D detectors. As demonstrated in [100, 101], the combination of high-energy X-rays (70–90 keV) and a large 2D detector makes it possible to col-

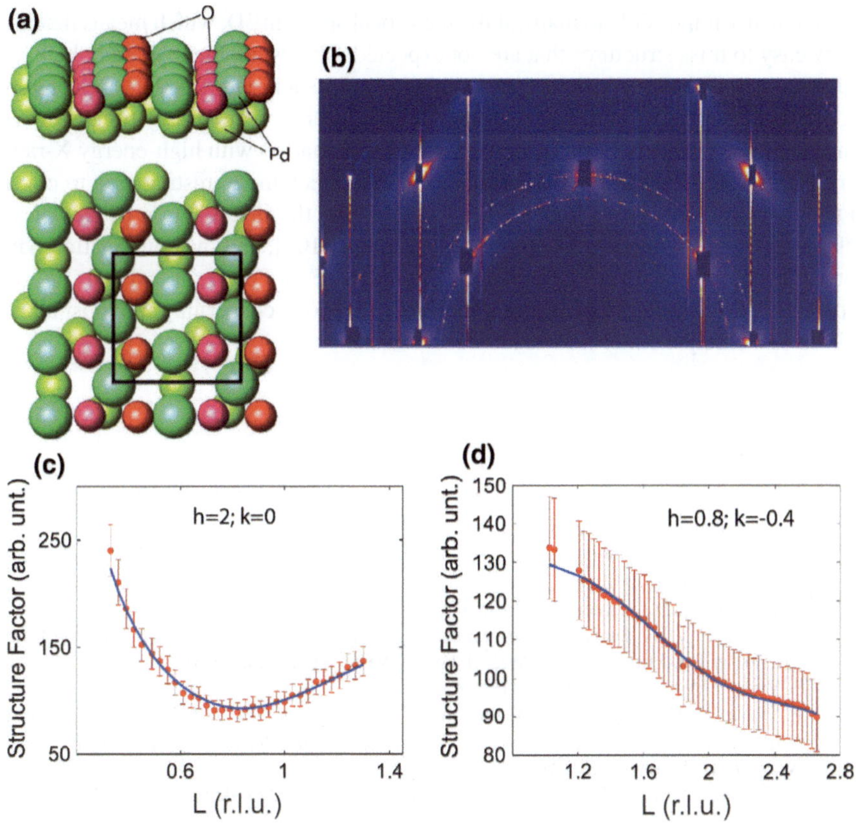

Fig. 3.11 In situ HESXRD data from the surface oxide on Pd(100) measured during CO oxidation in a flow of 6 Torr CO and 3 Torr O_2 at a sample temperature of 600 K. **a** Top and side view of the relevant oxygen-induced $(\sqrt{5} \times \sqrt{5})R27°$ surface oxide structure. The structure is an O-Pd-O trilayer corresponding to one PdO(101) plane. **b** All images collected during the rotational scan combined into a single image, in which the CTRs and superlattice rods can be directly seen. **c** and **d** Extracted (*dots*) CTRs and superlattice rods from the rotational images as seen in (**b**), and calculated structure factors (*full lines*). From [100]

lect from a single-crystal surface a large range of diffraction angles with a stationary detector, and through a simple rotation of the sample, a 3D map of the reciprocal space is acquired in about 15 min. The strength of the combination of high photon energies with 2D detectors for the investigation of epitaxial nanoparticle model catalyst systems was discussed in Sect. 3.3.3. As illustrated by Fig. 3.11 for the case of the Pd(100) surface during CO oxidation, the resulting data can be represented in several different ways in order to directly show the presence of different CTRs and superstructure rods, and to extract data for quantitative surface-structure determinations. These measurements are superior to conventional SXRD in the sense that they speed up the measurements by several orders of magnitude. In conventional SXRD it is practically impossible to map out the reciprocal space in 3D, which means that it is very easy to miss structures that are not expected. This is not the case with the large 2D detector. The large detector is also able to explore a higher fraction of the reciprocal space without rotating the sample, which improves the use of time-resolved measurements. This is especially true for the combination with high-energy X-rays, since the flatter Ewald sphere makes it easier to select an interesting area to cover. Future experiments using high energies will enable the investigation of nanoparticle model catalysts under flow reaction conditions [102], and the combination with other techniques such as laser induced fluorescence or Fourier transform infrared spectroscopy will shed new light on the open questions concerning the most active phase of a catalyst under flow reaction conditions.

References

1. R. Feidenhans'l, Surf. Sci. Rep. **10**, 105 (1989)
2. H. Dosch, *Critical phenomena at surfaces and interfaces*, Springer Tracts in Modern Physics. (Springer, Berlin, 1992), p. 126
3. I.K. Robinson, D.J. Tweet, Rep. Prog. Phys. **55**, 599 (1992)
4. A. Stierle, E. Vlieg, *Surface Sensitive X-ray Diffraction Methods in Modern Diffraction Methods*, ed. by E.J. Mittemeijer, U. Welzel. (Wiley VHC Weinheim, 2013)
5. E. Vlieg, J. Appl. Cryst. **30**, 532 (1997)
6. A. Stierle, H.-D. Carstanjen, W. Hofmann, O.H. Seeck, *Structural and Chemical Characterization on the Nanoscale in Nanoelectronics and Information Technology, edited by R* (Waser, Wiley VCH Weinheim, 2012)
7. N. Kasper, A. Stierle, P. Nolte, Y. Jin-Phillipp, T. Wagner, D.G. de Oteyza, H. Dosch, Surf. Sci. **600**, 2860 (2006)
8. U. Pietsch, V. Holy, T. Baumbach, *High-Resolution X-ray Scattering* (Springer, Berlin, 2004)
9. G. Renaud, R. Lazzari, F. Leroy, Surf. Sci. Rep. **64**, 255 (2009)
10. http://www.lightsources.org/
11. B.E. Warren, *X-ray Diffraction* (Dover Publications Inc, Mineola, 1990)
12. M. Lohmeier, E. Vlieg, J. Appl. Cryst. **26**, 706 (1993)
13. C.M. Schlepütz, R. Herger, P.R. Willmott, B.D. Patterson, O. Bunk, Ch. Brönnimann, B. Henrich, G. Hülsen, E.F. Eikenberry, Acta Cryst. **A61**, 418 (2005)
14. J. Drnec, T. Zhou, S. Pintea, W. Onderwaater, E. Vlieg, G. Renaud, R. Felicia, J. Appl. Cryst. **47**, 365 (2014)
15. D.K. Saldin, V.L. Shneerson, J. Phys.: Condens. Matter **20**, 304208 (2008)
16. L.G. Parratt, Phys. Rev. **95**, 359 (1954)
17. L. Nèvot, P. Croce, Revue de Physique Appl. **15**, 761 (1980)
18. W. Weber, B. Lengeler, Phys. Rev. B Rapid Comm. **46**, 7953 (1992)
19. S.K. Sinha, E.B. Sirota, S. Garoff, H.B. Stanley, Phys. Rev. B **38**, 2297 (1988)
20. J. Als-Nielsen, D. McMorrow, *Elements of Modern X-ray Physics* (Wiley, 2011)
21. O. Balmes, R. van Rijn, D. Wermeille, A. Resta, L. Petit, H. Isern, T. Dufrane, R. Felici, Catal. Today **145**, 220 (2009)
22. P. Bernard, K. Peters, J. Alvarez, S. Ferrer, Rev. Sci. Instrum. **70**, 1478 (1999)
23. R. van Rijn et al., Rev. Sci. Instrum. **81**, 014101 (2010)
24. M.-C. Saint-Lager, A. Bailly, P. Dolle, R. Baudoing-Savois, P. Taunier, S. Garaude, S. Cuccaro, S. Douillet, O. Geaymond, G. Perroux, O. Tissot, J.-S. Micha, O. Ulrich, F. Rieutord, Rev. Sci. Instrum. **78**, 083902 (2007)
25. A.K. Santra, D.W. Goodman, J. Phys.: Condens. Matter. **14**, R31 (2002)

26. H. Over, Y.D. Kim, A.P. Seitsonen, S. Wendt, E. Lundgren, M. Schmid, P. Varga, A. Morgante, G. Ertl, Science **287**, 1474 (2000)
27. A. Nagy, G. Mestl, Appl. Cat. A: General **188**, 337 (1999)
28. V.I. Bukhtiyarov, M. Hvecker, V.V. Kaichev, A. Knop-Gericke, R.W. Mayer, R. Schlögl, Phys. Rev. B **67**, 235422 (2003)
29. G. Rovida, F. Pratesi, M. Maglietta, E. Ferroni, Surf. Sci. **43**, 230 (1974)
30. C.I. Carlisle, D.A. King, M.-L. Bocquet, J. Cerda, P. Sautet, Phys. Rev. Lett. **84**, 3899 (2000)
31. W.-X. Li, C. Stampfl, M. Scheffler, Phys. Rev. B **67**, 045408 (2003)
32. M. Todorova, W.X. Li, M.V. Ganduglia-Pirovano, C. Stampfl, K. Reuter, M. Scheffler, Phys. Rev. Lett. **89**, 096103 (2002)
33. A. Michaelides, K. Reuter, M. Scheffler, J. Vac. Sci. Technol. A **23**, 1487 (2005)
34. M. Schmid, A. Reicho, A. Stierle, I. Costina, J. Klikovits, P. Kostelnik, O. Dubay, G. Kresse, J. Gustafson, E. Lundgren, J.N. Andersen, H. Dosch, P. Varga, Phys. Rev. Lett. **96**, 146102-1-4 (2006)
35. J. Schnadt, A. Michaelides, J. Knudsen, R.T. Vang, K. Reuter, E. Lægsgaard, M. Scheffler, F. Besenbacher, Phys. Rev. Lett. **96**, 146101 (2006)
36. A. Reicho, A. Stierle, I. Costina, H. Dosch, Surf. Sci. **601** L19L23 (2007)
37. A. Stierle, I. Costina, S. Kumaragurubaran, H. Dosch, J. Phys. Chem. C **111**, 10998 (2007)
38. M. Rocca, L. Savio, L. Vattuone, U. Burghaus, V. Palomba, N. Novelli, F. Buatier de Mongeot, U. Valbusa, R. Gunnella, G. Comelli, A. Baraldi, S. Lizzit, G. Paolucci, Phys. Rev. B **61**, 213 (2000)
39. H. Iddir, D.D. Fong, P. Zapo, P.H. Fuoss, L.A. Curtiss, G.-W. Zhou, J.A. Eastman, Phys. Rev. B **76**, 241404 R (2007)
40. E. Lundgren, G. Kresse, C. Klein, M. Borg, J.N. Andersen, M. De Santis, Y. Gauthier, C. Konvicka, M. Schmid, P. Varga, Phys. Rev. Lett. **88**, 246103 (2002)
41. P. Kostelnik, N. Seriani, G. Kresse, A. Mikkelsen, E. Lundgren, V. Blum, T. Sikola, P. Varga, M. Schmid, Surf. Sci. **601**, 1574 (2007)
42. B.L.M. Hendriksen, S.C. Bobaru, J.W.M. Frenken, Surf. Sci. **552**, 229 (2004)
43. J. Rogal, K. Reuter, M. Scheffler, Phys. Rev. Lett. **98**, 046101 (2007)
44. M.S. Chen, Y. Cai, Z. Yan, K.K. Gath, S. Axnanda, D.W. Goodman, Surf. Sci. **601**, 5326 (2007)
45. J. Gustafson, A. Mikkelsen, M. Borg, E. Lundgren, L. Köhler, G. Kresse, M. Schmid, P. Varga, J. Yuhara, X. Torrelles, C. Quiros, J.N. Andersen, Phys. Rev. Lett. **92**, 126102 (2004)
46. E. Lundgren, J. Gustafson, A. Mikkelsen, J.N. Andersen, A. Stierle, H. Dosch, M. Todorova, J. Rogal, K. Reuter, M. Scheffler, Phys. Rev. Lett. **92**, 046101 (2004)
47. A. Stierle, N. Kasper, H. Dosch, E. Lundgren, J. Gustafson, A. Mikkelsen, J.N. Andersen, J. Chem. Phys. **122**, 044706 (2005)
48. N. Kasper, P. Nolte, A. Stierle, J. Phys. Chem. C **116**(40), 21459 (2012)
49. R. Westerström, C.J. Weststrate, J. Gustafson, A. Mikkelsen, J. Schnadt, J.N. Andersen, E. Lundgren, N. Seriani, F. Mittendorfer, G. Kresse, A. Stierle, Phys. Rev. B **80**, 125431 (2009)
50. C.W. Tucker, J. Appl. Phys. **37**, 3013 (1966)
51. J. Gustafson, A. Mikkelsen, M. Borg, J.N. Andersen, E. Lundgren, C. Klein, W. Hofer, M. Schmid, P. Varga, L. Köhler, G. Kresse, N. Kasper, A. Stierle, H. Dosch, Phys. Rev. B **71**, 115442 (2005)
52. C. Dri, C. Africh, F. Esch, G. Comelli, O. Dubay, L. Köhler, F. Mittendorfer, G. Kresse, P. Dudin, M. Kiskinova, J. Chem. Phys. **125**, 094701 (2006)
53. W.X. Li, L. österlund, E.K. Vestergaard, R.T. Vang, J. Matthiesen, T.M. Pedersen, E. Lægsgaard, B. Hammer, F. Besenbacher, Phys. Rev. Lett. **93**, 146104 (2004)
54. M.D. Ackermann, T.M. Pedersen, B.L.M. Hendriksen, O. Robach, S.C. Bobaru, I. Popa, C. Quiros, H. Kim, B. Hammer, S. Ferrer, J.W.M. Frenken, Phys. Rev. Lett. **95**, 255505 (2005)
55. C. Ellinger, A. Stierle, I.K. Robinson, A. Nefedov, H. Dosch, J. Phys. Cond. Mat. **20**, 184013 (2008)
56. Y.B. He, A. Stierle, W.X. Li, A. Farkas, N. Kasper, H. Over, J. Phys. Chem. C **112**, 11946 (2008)

57. R. Westerström, J. Gustafson, A. Resta, A. Mikkelsen, J.N. Andersen, E. Lundgren, N. Seriani, F. Mittendorfer, M. Schmid, J. Klikovits, P. Varga, M.D. Ackermann, J.W.M. Frenken, N. Kasper, A. Stierle, Phys. Rev. B **76**, 155410 (2007)
58. A. Vlad, A. Stierle, R. Westerström, S. Blomberg, A. Mikkelsen, E. Lundgren, Phys. Rev. B **86**, 035407 (2012)
59. J. Gustafson, A. Resta, A. Mikkelsen, R. Westerström, J.N. Andersen, E. Lundgren, J. Weissenrieder, M. Schmid, P. Varga, N. Kasper, X. Torrelles, S. Ferrer, F. Mittendorfer, G. Kresse, Phys. Rev. B **74**, 035401 (2006)
60. M.A. Newton, C. Belver-Coldeira, A. Martinez-Arias, M. Fernandez-Garcia, Nat. Mat. **6**, 528 (2007)
61. C.R. Henry, Surf. Sci. Rep. **31**, 231 (1998)
62. P. Nolte, A. Stierle, N. Kasper, N. Jeutter, H. Dosch, Nano Lett. **11**, 4697 (2011)
63. P. Nolte, A. Stierle, N.Y. Jin-Phillipp, N. Kasper, T.U. Schulli, H. Dosch, Science **321**, 1654–1658 (2008)
64. F. Mittendorfer, N. Seriani, O. Dubay, G. Kresse, Phys. Rev. B **76**, 233413 (2007)
65. P. Nolte, A. Stierle, N. Kasper, N.Y. Jin-Phillipp, H. Reichert, A. Rühm, J. Okasinski, H. Dosch, S. Schöder, Phys. Rev. B **77**, 115444 (2008)
66. P. Müller, U. Hejral, U. Rütt, A. Stierle, Phys. Chem. Chem. Phys. **16**, 13866 (2014)
67. U. Hejral, A. Vlad, P. Nolte, A. Stierle, J. Phys. Chem. C **117**(39), 19955 (2013)
68. K.F. Peters, C.J. Walker, P. Steadman, O. Robach, H. Isern, S. Ferrer, Phys. Rev. Lett. **86**, 5325 (2001)
69. M.C. Saint-Lager, Y. Jugnet, P. Dolle, L. Piccolo, R. Baudoing-Savois, J.C. Bertolini, A. Bailly, O. Robach, C. Walker, S. Ferrer, Surf. Sci. **587**, 229 (2005)
70. M.D. Ackermann, T.M. Pedersen, B.L.M. Hendriksen, O. Robach, S.C. Bobaru, I. Popa, C. Quiros, H. Kim, B. Hammer, S. Ferrer, J.W.M. Frenken, Phys. Rev. Lett. **95**, 255505 (2005)
71. Y.B. He, M. Knapp, E. Lundgren, H. Over, J. Phys. Chem. B **109**, 21825 (2005)
72. H. Over, O. Balmes, E. Lundgren, Surf. Sci. **603**, 298 (2009)
73. H. Over, O. Balmes, E. Lundgren, Catal. Today **145**, 236 (2009)
74. S. Zweidinger, D. Crihan, M. Knapp, J.P. Hofmann, A.P. Seitsonen, C.J. Westrate, E. Lundgren, J.N. Andersen, H. Over, J. Phys. Chem. C **112**, 9966 (2008)
75. H. Deacon, US Patent (1875) 0165802
76. K. Iwanaga, K. Seki, T. Hibi, K. Issoh, T. Suzuta, M. Nakada, Y. Mori, T. Abe, The development of improved hydrogen chloride oxidation process. Sumitomo Kagaku **I**, 1 (2004)
77. D. Crihan, M. Knapp, S. Zweidinger, E. Lundgren, C.J. Weststrate, J.N. Andersen, A.P. Seitsonen, H. Over, Angew. Chem. Int. Ed. **47**, 2131 (2008)
78. N. López, J. Gmez-Segura, R.P. Marn, J. Prez-Ramrez, J. Catal. **255**, 29 (2008)
79. S. Zweidinger, J.P. Hofmann, O. Balmes, E. Lundgren, H. Over, J. Catal. **272**, 169 (2010)
80. M.V. Ganduglia-Pirovano, M. Scheffler, A. Baraldi, S. Lizzit, G. Comelli, G. Paolucci, R. Rosei, Phys. Rev. B **63**, 205415 (2001)
81. A. Baraldi, J. Phys. Condens. Matter **20**, 093001 (2008)
82. C.H.F. Peden, D.W. Goodman, D.S. Blair, P.J. Berlowitz, G.B. Fischer, S.H. Oh, J. Phys. Chem. **92**, 1563 (1988)
83. J. Gustafson, R. Westerström, A. Mikkelsen, X. Torrelles, O. Balmes, J.N. Andersen, C.J. Baddeley, E. Lundgren, Phys. Rev. B **78**, 045423 (2008)
84. J. Gustafson, R. Westerström, A. Mikkelsen, A. Resta, J.N. Andersen, X. Torrelles, O. Balmes, M. Schmid, P. Varga, B. Hammer, G. Kresse, C.J. Baddeley, E. Lundgren, Catal. Today **145**, 227 (2009)
85. J. Klikovits, M. Schmid, L.R. Merte, P, Varga, R. Westerström, A. Resta, J.N. Andersen, J. Gustafson, A. Mikkelsen, E. Lundgren, F. Mittendorfer, G. Kresse, Phys. Rev. Lett. **101**, 266104 (2008)
86. R. Westerström, J.G. Wang, M. Ackermann, J. Gustafson, A. Resta, A. Mikkelsen, J.N. Andersen, E. Lundgren, O. Balmes, X. Torrelles, J.W.M. Frenken, B. Hammer, J. Phys. Condens. Matter **20**, 184018 (2008)
87. G. Rupprechter, K. Hayek, H. Hofmeister, J. Catal. **173**, 409 (1998)

88. S. Blomberg, R. Westerström, N.M. Martin, E. Lundgren, J.N. Andersen, M.E. Messing, J. Gustafson, Surf. Sci. **628**, 153 (2014)
89. I. Laoufi, M.-C. Saint-Lager, R. Lazzari, J. Jupille, O. Robach, S. Garaudee, G. Cabailh, P. Dolle, H. Cruguel, A. Bailly, J. Phys. Chem. C **4673**, 115 (2011)
90. P. Nolte, A. Stierle, O. Balmes, V. Srot, P.A. van Aken, L.P.H. Jeurgens, H. Dosch, Catal. Today **145**, 243 (2009)
91. R. van Rijn, O. Balmes, R. Felici, J. Gustafson, D. Wermeille, R. Westerström, E. Lundgren, J.W.M. Frenken, J. Phys. Chem. C **114**, 6875–6876 (2010)
92. R. van Rijn, O. Balmes, A. Resta, D. Wermeille, R. Westerström, J. Gustafson, R. Felici, E. Lundgren, J.W.M. Frenken, Phys. Chem. Chem. Phys. **13**, 13167–13171 (2011)
93. F. Gao, Y. Wang, Y. Cai, D.W. Goodman, J. Phys. Chem. C **113**, 174–181 (2009)
94. B.L.M. Hendriksen, M.D. Ackermann, R. van Rijn, D. Stoltz, I. Popa, O. Balmes, A. Resta, D. Wermeille, R. Felici, S. Ferrer, J.W.M. Frenken, Nat. Chem. **2**, 730–734 (2010)
95. J. Gustafson, R. Westerström, O. Balmes, A. Resta, R. van Rijn, X. Torrelles, C.T. Herbschleb, J.W.M. Frenken, E. Lundgren, J. Phys. Chem. C **114**, 4580–4583 (2010)
96. J. Gustafson, R. Westerström, O. Balmes, A. Resta, R. van Rijn, X. Torrelles, C.T. Herbschleb, J.W.M. Frenken, E. Lundgren, J. Phys. Chem. C **114**, 22372–22373 (2010)
97. O. Balmes, G. Prevot, X. Torrelles, E. Lundgren, S. Ferrer, J. Catal. **309**, 33–37 (2014)
98. C. Zhang, E. Lundgren, P.-A. Carlsson, O. Balmes, A. Hellman, L.R. Merte, M. Shipilin, W. Onderwaater, J. Gustafson, J. Phys. Chem. C **119**, 11646–11652 (2015)
99. A. Hellman, A. Resta, N.M. Martin, J. Gustafson, A. Trinchero, P.-A. Carlsson, O. Balmes, R. Felici, R. van Rijn, J.W.M. Frenken, J.N. Andersen, E. Lundgren, H. Grönbeck, J. Phys. Chem. Lett. **3**, 678–682 (2012)
100. J. Gustafson, M. Shipilin, C. Zhang, A. Stierle, U. Hejral, U. Ruett, O. Gutowski, P.-A. Carlsson, M. Skoglundh, E. Lundgren, Science **343**, 758–761 (2014)
101. M. Shipilin, U. Hejral, E. Lundgren, L.R. Merte, C. Zhang, A. Stierle, U. Ruett, O. Gutowski, M. Skoglundh, P.-A. Carlsson, J. Gustafson, Surf. Sci. **630**, 229–235 (2014)
102. U. Hejral, P. Müller, O. Balmes, D. Pontoni, A. Stierle, Nat. Comm. **7**, 10964 (2016)

Chapter 4
From Spectator Species to Active Site Using X-ray Absorption and Emission Spectroscopy Under Realistic Conditions

Maarten Nachtegaal, Urs Hartfelder and Jeroen A. van Bokhoven

Abstract X-ray absorption and emission spectroscopy experiments on metal-supported catalysts are described. Time- and space-resolved spectroscopy measurements enable determining the electronic and geometric structures of the catalytically active elements at any location within a real reactor and under actual reaction conditions. Methods that distinguish spectator from active species are described. The data survey shows that there is not one single phase that is responsible for activity. Conversion leads to changes in gas composition, affecting the structure. Within a single reactor, multiple phases can be present, which is certainly the case under oscillating conditions. Platinum catalysts contain lowly-active phases that are associated with carbon monoxide poisoning. Highly-active phases form after complete desorption of carbon monoxide. Such a surface is also prone to very rapid surface oxidation. The relationship between these phases and catalytic activity in carbon monoxide oxidation is discussed.

4.1 Introduction

In situ and *operando* catalyst characterization methods using X-ray absorption and emission spectroscopy (XAS and XES) are indispensable tools to determine the structure of catalytically-active sites under working conditions [1, 2]. By using these techniques, the local electronic and geometric structure (up to ca. 6 Å) around

M. Nachtegaal
Paul Scherrer Institute, 5236 Villigen, Switzerland
e-mail: maarten.nachtegaal@psi.ch

U. Hartfelder · J.A. van Bokhoven (✉)
Institute for Chemical and Bioengineering, ETH Zurich, 8093 Zurich, Switzerland
e-mail: jeroen.vanbokhoven@chem.ethz.ch

U. Hartfelder
e-mail: urs.hartfelder@chem.ethz.ch

J.A. van Bokhoven
Paul Scherrer Institute, 5236 Villigen, Switzerland

J. Frenken and I. Groot (eds.), Operando *Research in Heterogeneous Catalysis*,
Springer Series in Chemical Physics 114, DOI 10.1007/978-3-319-44439-0_4

the atom of interest can be determined. X-rays of sufficiently high energy (4 keV or higher) have a high penetration depth, so that catalysts in an actual reactor can be measured and the catalyst structure under reaction conditions can be accessed in an element-specific manner. Because of the strong interaction of reactants and intermediates with the catalyst surface, the catalyst structure varies with the conditions. To perform time-resolved XAS or XES experiments, a high flux of X-rays is needed, which is provided by synchrotron light sources. In general, each synchrotron has at least one beam line dedicated to X-ray spectroscopy, many of which are dedicated to catalysis research [3].

In an X-ray absorption experiment, a monochromatic X-ray beam is scanned around the absorption edge of the element of interest. A core electron is excited by a photon when the energy of the incoming photon matches or exceeds the binding energy of that core electron. At that energy, a sudden jump in absorption occurs (Fig. 4.1). The exceeding energy of the incoming photon is given to the outgoing photon as kinetic energy. Because the absorption coefficient is a function of the X-ray energy, the absorption signal is modulated. Just above the absorption edge, the outgoing photon does not contain enough kinetic energy to be excited into the vacuum and instead could be excited to unoccupied electronic states. In case of excitation of an electron in the K shell, a 1s electron is excited into empty p density of states, which is the dipolar transition, respectively empty d density of states, the quadrupolar transition. Exciting an electron from the L shell probes p density of states (L_1 edge, 2s electron excitation), respectively d density of states ($L_{2,3}$ edges, 2p excitation). From these considerations, the following becomes clear: i. XAS is element specific, because of the unique binding energy of each element's electron; ii. X-ray absorption K and L_1 edge spectra measure the empty p density of states and are thus sensitive to geometry and oxidation state; iii. $L_{2,3}$ edge spectra probe empty d density of states. In case of K edge absorption, the weak signals of the

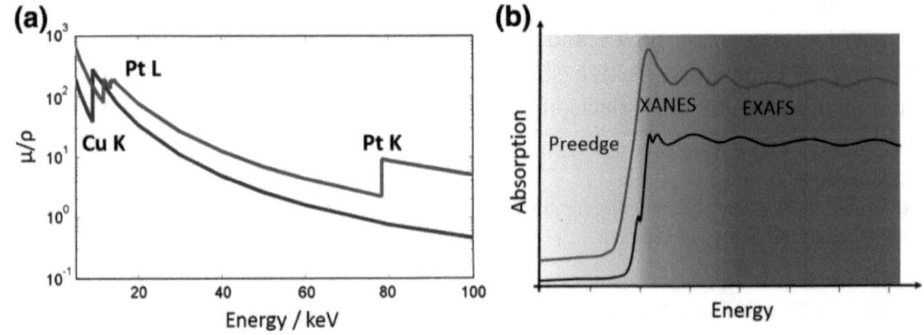

Fig. 4.1 Transmission of X-rays through matter as function of X-ray energy. Sharp rises in absorption occur at energies equal to the binding energy of electrons (**a**). A XAS spectrum is typically divided into three regions, the *pre-edge*, the *XANES*, and the *EXAFS* regions (**b**). The *pre-edge* and *XANES* are separated by the absorption edge, characterized by the sharp increase in X-ray absorption. The *bottom line* corresponds to a Cu K-edge spectrum, the *top line* to a Pt L_3-edge XAS spectrum

quadrupolar transition probe d density of states. When measuring transition metals, measuring the L_3 edge may be attractive, because it tracks the anti-bonding states of adsorbates on the surface and is sensitive to the mode of adsorption in a quantitative manner (Fig. 4.2) [4]. In summary, the X-ray absorption near edge region (XANES) provides element-specific information on the empty density of states.

Observing the modulation of absorption over a much longer energy interval yields EXAFS data, which probe the local environment around the X-ray absorbing atom. The averaged coordination, the distance to the near neighbors, and the disorder (Debye-Waller factor) can be obtained from fitting of the EXAFS spectra [5].

In addition to X-ray absorption, X-ray emission yields complementary information. An X-ray emission spectrum tracks the emitted fluorescence upon elimination of the core-hole generated by the photon absorption. This experiment probes the filled density of states and is thus a complimentary tool to determine the electronic structure. Because the type of neighbors and oxidation state determine the electronic structure X-ray emission spectroscopy is a sensitive tool for their determination [6–8].

In situ cells that enable measuring X-ray absorption and X-ray emission spectroscopy on working catalysts have been around for many decades [9]. They can be constructed because the X-rays of sufficiently high energy used for X-ray absorption and emission spectroscopy penetrate through thin reactor walls. Heterogeneous and homogeneous reactions can be accessed at high pressure (in excess of 100 s bar) and temperature (beyond 1000 °C) [10]. Such measurements yield the structure of the functioning catalyst, sometimes performed under industrial conditions. A complication of the high penetration depth of X-rays is that an average of all structures of the measured element is yielded. Thus, in case the catalytically-active site relates to a minority of all species within the sample, it is difficult to extract the exact structure of the site. However, because X-ray absorption and emission

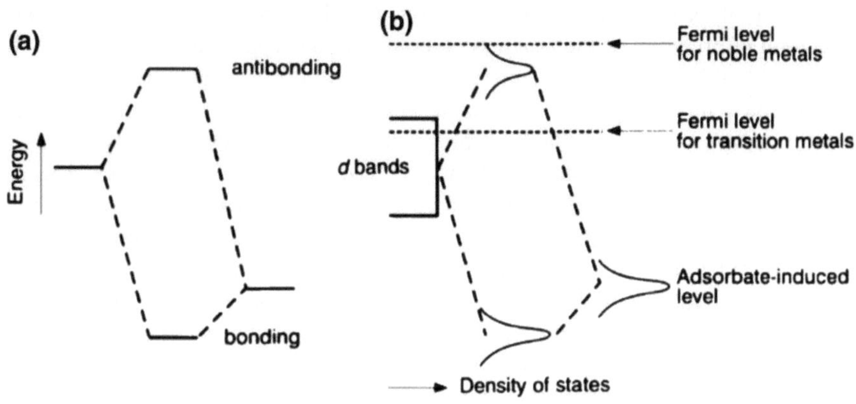

Fig. 4.2 Formation of bonding and anti-bonding states upon bonding of an adsorbate **a** with an atom and **b** on a metal surface. XAS probes the empty (anti-bonding) states and XES the filled (bonding and non-bonding) states. Reproduced with permission from [50]

spectroscopy is particularly sensitive to differences between spectra, modulation of conditions and analyzing the changes in the spectra yields the structure of the sites that respond to the modulation in the conditions [11–13]. Transient and modulation-excited spectroscopy coupled with time-resolved measuring enables the selective removal of the spectator species signals [14]. Thus an enormous enhancement of sensitivity is obtained.

During a catalytic cycle, the structure of a catalyst changes. After reactant adsorption, a chemisorbed intermediate forms, which causes a change in structure and possibly oxidation state of some of the atoms making up the catalytically-active site; Then, the surface reaction occurs, which changes the adsorbed intermediate and with that possibly the site structure; Finally desorption of the product yields the original site free of adsorbates. During a steady-state experiment, the time-averaged structure of the site will be probed, which is dominated by the resting state of the catalyst. To enable following the evolution of catalyst structure, time-resolved measurements are essential as will be highlighted in this chapter.

This chapter features selected in situ and *operando* XAS and XES experiments that relate to carbon monoxide oxidation over platinum catalysts. It is highlighted how catalyst structure and structural changes can be deduced from static and transient experimental spectra. The vast majority of atoms within a catalyst do not contribute to catalytically-active sites.

4.2 Transient X-ray Absorption Spectroscopy

Time-resolved measurements are crucial for in situ studies of dynamic chemical systems in non-steady state environments, such as the elucidation of the structure of intermediates involved in chemical processes or the kinetics of transient phenomena in the structure of the catalyst and the catalyst-reactant interface. Traditionally, the collection of XAS spectra has been slow (10–15 min per scan) so that only structural information of the progress of slow reactions could be obtained. With the advent of highly brilliant third generation synchrotron sources and the development of dedicated monochromators, detectors and data acquisition systems, structural changes that occur during reactions proceeding on timescales of (sub) seconds to several minutes can be followed. Time resolved XAS (quick XAS [15, 16] and energy dispersive XAS [17, 18]), with an ultimate time resolution of a few milliseconds per spectrum, allow to identify the structure of intermediates in catalytic reactions. When coupled with techniques to determine the performance/activity of the catalyst, for example mass spectrometry or gas chromatography, true structure-performance relationships can be obtained. However, as pointed out by Bravo-Suárez et al. [19], the sole identification of the presence of intermediate species is not sufficient proof that they are truly reaction intermediates; They could be spectator species during the actual reaction. Oyama and coworkers showed the value of transient XAS for the identification of active sites by studying the epoxidation of propylene by gold catalysts: They showed that the rate of structural change of the titanium site corresponds to the observed global reaction rate [19]. This was one of the first reports that showed the strength of transient XAS, in which solid-state kinetic estimations were carried out by XAS spectroscopy and were directly coupled to the global reaction rate to determine if intermediates were involved in the reaction.

Gas phase reactions in an in situ catalytic cell are typically triggered by a fast exchange of reactive gases. By using gas lines and capillaries with small inner diameters and by positioning the capillaries close to the fast switching valves, reactions at the beginning of the reactor bed can be triggered within a few milliseconds. The changes then typically happen within a few tens of milliseconds up to a few seconds. Thus, in order to detect changes well, the detection limit for spectroscopy ultimately needs to be a few milliseconds. Currently there are two fundamental different approaches that reach this time resolution, each with its advantages and disadvantages. Both these techniques need very powerful X-ray sources, so-called third generation synchrotron sources, which provide at least 10^{12} monochromatic photons per second.

With QuickXAS, the monochromatic beam is selected from the polychromatic synchrotron beam by the use of a channel-cut crystal (i.e. two well defined crystal surfaces carved out in one crystal block). The first crystal selects out the energy from the polychromatic beam according to Bragg's law. The second crystal is needed to maintain a fixed exit of the beam and thus a stable beam position at the sample. This channel-cut crystal is then oscillated back and forth, i.e. moved through an energy range of a few 100 eV up to a few keV. Currently, mechanics allow for an oscillation of 50 Hz, the limitation is rather the speed of the detection system. A time resolution is achieved on the order of 10 ms for a full EXAFS spectrum (1000 eV), which is what one ultimately needs to follow solid-gas interface reactions in a capillary reactor. The main advantage of QEXAFS is that it is easily integrated into any existing XAS beamline.

The energy-dispersive XAFS technique has preceded the developments of QEXAFS and has meanwhile matured at a few synchrotron sources. Nowadays one obtains a milli-second time resolution with this technique. In energy dispersive XAS a curved crystal polychromator disperses and focuses a polychromatic X-ray beam onto the sample. The transmitted beam is then detected by a position-sensitive detector. The ultimate time resolution is only determined by the readout speed of the detector and the number of photons available in such a small time scale. EDXAS requires a very stable beamline setup and is thus available on a few dedicated beamlines.

The high time-resolution of XAS combined with modulated excitation offers the possibility to pick up small structural changes in a catalyst mediated through the modulation excitation. In a modulated excitation experiment, a material is excited with a periodically alternating external stimulation such as temperature, pressure, magnetic field, pH, potential, or concentration. At the same time the response of the catalyst is monitored with time-resolved spectroscopy. One can then analyze the large time-resolved data set either with multivariant analysis tools or with phase-sensitive detection. In phase-sensitive detection (PSD), the measured system response towards the modulated excitation, in our case the set of XAS spectra, is filtered with the excitation frequency. Processes with periodically repeated oxidative regeneration are prime examples of such a modulation experiment, where the material is periodically exposed to alternating gas compositions. Mathematically phase-sensitive detection is described by the following equation [20]:

$$A_k^{\phi_k^{PSD}}(E) = \frac{2}{T}\int_0^T A(E,t)\sin(k\omega t + \phi_k^{PSD})dt \tag{4.1}$$

where $A(E, t)$ is the original signal (i.e. the XAS spectrum) as a function of time t and energy E, $k = 1$ is the fundamental harmonic, ω is the frequency of the external stimulation, ϕ_k^{PSD} is the demodulation phase angle, and $T = 1/\omega$ is the modulation period. Recent modulated excitation XAS experiments showed the feasibility of recording demodulated XAS spectra [11, 21, 22]. These studies demonstrated that minority species can be detected in modulation excitation XAS and that noise and spectator species are removed from the demodulated XAS spectra. By calculating the demodulated spectrum $A(E)$ for phase angles between 0 and 360°, the spectra are transformed from the time-domain into a phase-domain. This mathematical procedure cancels out all parts of the measured spectra that do not follow the excitation frequency ω. These parts include the spectator species and noise. Thus, demodulated spectra contain only the differences between different phase angles of the modulation period. In other words, PSD is sensitive only to the changes imposed by the external stimulus. When the XAS spectra only show two distinct states, the demodulated spectra will show the same spectral features as the difference spectrum of spectra of these two distinct states, only with much less noise [12]. The modulated excitation approach has further been applied to infrared spectroscopy [20, 23, 24], Raman spectroscopy, and to X-ray diffraction (XRD) where a correlated temporal analysis of the Raman and the XRD signals allowed a more detailed study of phase transitions [25, 26].

An alternative to phase sensitive detection to analyze the modulation-excitation XAS data set, to find the structure of a minority (active) species, is to use multivariate analysis software. This was applied recently to time-resolved XAS spectra and has great potential for determining the structure of either short-lived intermediates or of minority species [27, 28]. This approach has as additional benefit that the time evolution of the different species present is preserved.

4.3 X-ray Emission Spectroscopy

X-ray emission spectroscopy (XES) is complementary to XAS as it probes the occupied density of states around the atom of interest. XES includes a range of hard-X-ray (>5 keV) photon-in-photon-out techniques, including the analysis of the fine structure of the X-ray emission lines and resonant inelastic X-ray scattering (RIXS) [29, 30]. In comparison to conventional X-ray absorption techniques, RIXS provides an enhanced sensitivity to the electronic structure where both the occupied and unoccupied electronic states are simultaneously probed. In a RIXS experiment the incoming photon (photon-in) excites an inner-shell electron just above the Fermi level. By tuning the incident beam energy, different unoccupied states can be probed. The remaining core-hole is filled within a few femtoseconds [31] by electronic transitions from higher shells of the atom. This electronic transition is accompanied by X-ray emission (photon-out), the energy of which depends on the electronic states involved in the decay transition and thus provides information on the occupied electronic states.

RIXS spectroscopy can be employed for the detection of electronic transitions involving core (core-to-core RIXS) or valence electrons (valence-to-core RIXS). In case of core-to-core RIXS, the initial and final electronic states of an atom are in most cases screened and therefore are rather insensitive to its chemical surrounding. For this reason, the experimental results of core-to-core RIXS can be evaluated straightforward by theoretical calculations that are based on the single-electron approximation (using for example the FEFF code [32, 33]). Moreover, these transitions provide highest X-ray rates (i.e. detected intensities) and therefore are commonly used for the detection of elements present at low concentrations.

In valence-to-core RIXS, the inner core-hole is filled by an outer-shell electron. Because of electron-electron interaction between the excited core-electron and the decaying valence-electron, detailed information about the electronic states of an atom can be obtained. In valence-to-core RIXS spectroscopy, both the incoming and outgoing photon energies are in the keV range while the probed unoccupied and occupied electronic states are only in a range of a few eV around the Fermi level. Therefore, a picture of the electronic states of the studied system can be retrieved under in situ conditions, which is challenging for electron-based techniques such as X-ray photoelectron spectroscopy (XPS). The transition rates for valence-to-core RIXS are weaker as compared to core-to-core RIXS, therefore the experiments are time consuming and usually performed on concentrated samples.

In Fig. 4.3 an exemplary $L\alpha_1$-RIXS plane of Au_2O_3 is shown that was collected at room temperature around the Au L_3 edge by detecting the $L\alpha_1$ X-ray emission resulting from the $3d_{5/2} \rightarrow 2p_{3/2}$ core-to-core transition. At an incident energy of 11920.5 eV and emission energy of 9713.5 eV a strong $2p_{3/2} \rightarrow 5d$ resonance is detected. In Fig. 4.3 (right panel), two XAS spectra extracted from the RIXS plane are plotted. By integrating the full X-ray emission intensity at each incident energy, the total fluorescence yield (TFY) spectrum was constructed, while high resolution XAS (HR-XAS) is obtained from the maximum of the $L\alpha_1$ peak intensity. The HR-XAS exhibits more pronounced features, and therefore detailed information about the 5d electronic structure can be obtained [34].

The main advantage of RIXS is that it provides a full electronic picture of the metal site of interest, under in situ conditions. Similar high-resolution XAS spectra as extracted from the RIXS planes in Fig. 4.3 can also be collected by tuning a spectrometer to the top of the emission line and varying the incident energy. Such spectra are called high-energy-resolution fluorescence-detected XAS (HERFD XAS) spectra. They have better energy resolution than regular XAS spectra and thus reflect the density of states (DOS) in more detail. This has been used for

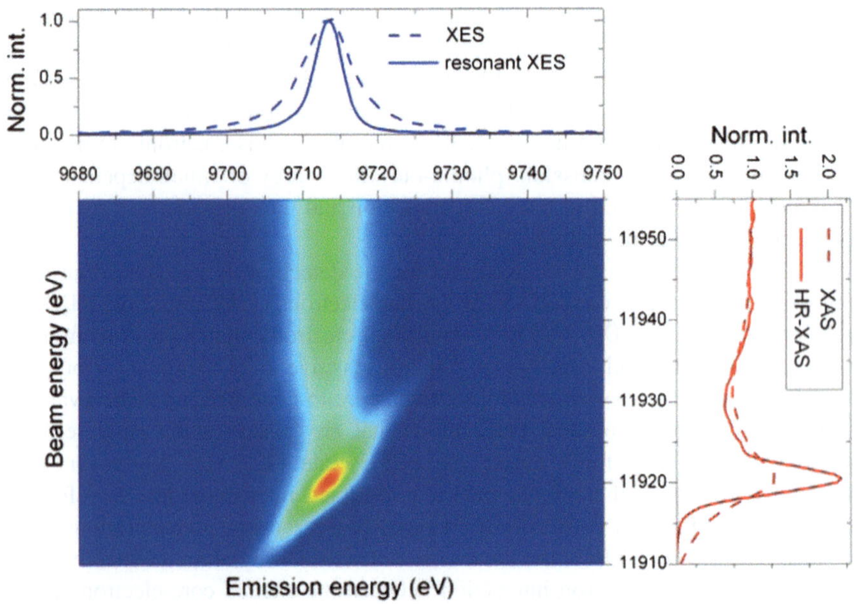

Fig. 4.3 Lα_1-RIXS plane of Au_2O_3 around the Au L_3 absorption edge. *Right* the TFY and HR-XAS spectra. *Top* resonant and non-resonant XES spectra. Reproduced with permission from [52]

example to explore the Cu K-edge XANES region of copper-loaded zeolites under in situ conditions [35, 36] (Fig. 4.4) to follow the changes in the copper sites of Cu-MOR (a medium pore-size zeolite) during different stages of partial oxidation of methane. The features in the pre-edge region of the spectra can be explained by the contribution of two oxidation states. Other features result from shake-up and shake-down processes which depend on the ligand and on the structure of the copper centers in the zeolite [36]. These studies of the unoccupied DOS provide useful insight in the reaction mechanism and the number of copper sites involved in the reaction, which turned out to be significantly larger than previously thought based on extraction studies [36].

The Kβ main line spectra are dominated by the valence shell spin state contributions and thus enable the detection of oxidation state changes and high-spin low-spin transitions. The spin sensitivity arises from an intra-atomic interaction between the 3p and 3d shells [6]. The satellite lines, i.e. the emission lines just below the Fermi level or alternatively the valence-to-core transitions (denoted Kβ'' or K$\beta_{2,5}$), show great sensitivity to changes in the chemical environment, such as ligand identity, ligand ionization state, and hybridization state [37]. The formal electronic configuration in the final state of the X-ray emission transition is identical to valence electron photoemission, such as X-ray photoelectron spectroscopy. However, there are some important differences between XPS and XES [38]. XES is a strictly element-specific probe; Only the electron density of the absorbing atom

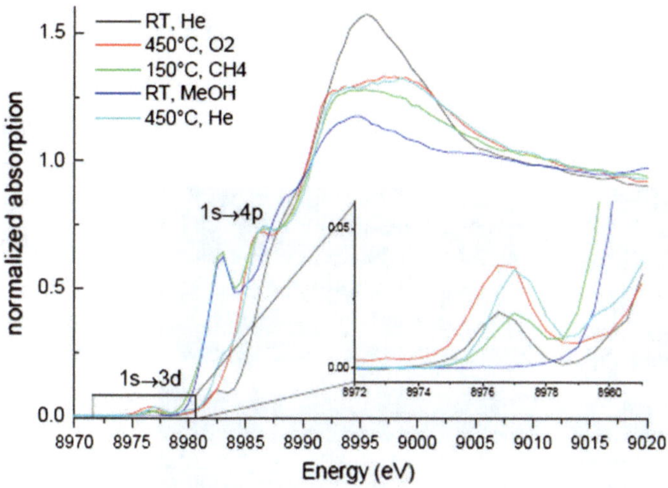

Fig. 4.4 Cu K edge HERFD XAS spectra of Cu-zeolites under different gas atmospheres show rich pre-edge features. Copper reduces from Cu^{2+} to a mixture of Cu^{+} and Cu^{2+} states upon interaction with methane. Shake-up and shake-down processes are also present in the pre-edge region [36]. Reproduced with permission from [36]

and its direct environment is probed with a considerably simpler analysis than for XPS experiments. The selection rules for electron transitions in K-edge XES only allow for orbitals to be detected that have p-contribution with respect to the metal atom. Furthermore, XES probes the bulk and not only the surface of the sample and vacuum conditions are not required.

Valence-to-core RIXS spectroscopy enables to probe the electronic structure of the shared electrons of the ligand environment of a metal of interest [37, 39, 40]. Since this is a very 'photon-hungry' technique, time-resolved experiments are limited, and only a static picture of the electronic state can be obtained. As an example, the unoccupied electronic states of titania (anatase) were probed by exciting the 1s electron above the Fermi level and probing the occupied electronic states by detecting the following 3p → 1s and valence → 1s decay transitions [41]. The recorded RIXS plane of titania is plotted in Fig. 4.5. The RIXS plane consists of well-separated pre-edge structures at beam energies between 4968 and 4975 eV and post-edge features at higher excitation energies.

Based on the HR-XAS and non-resonant XES spectra, the valence and conduction band electronic states can be extracted from the measured RIXS plane. The curves were scaled to the Fermi energy by shifting the energy axis of each spectrum by a value of 4965.4 eV. This energy was extracted from the inflection point of the valence-to-core transition (dashed line in Fig. 4.6). Figure 4.6 plots the recalibrated and renormalized spectra, which are compared to the density of states calculated with the FEFF8.4 code [33]. The titanium and oxygen DOS for Ti are plotted in Fig. 4.6 (bottom). The DOS computation shows that the pre-edge structure mainly consists of Ti d-states and the contribution of the oxygen p-states is negligibly

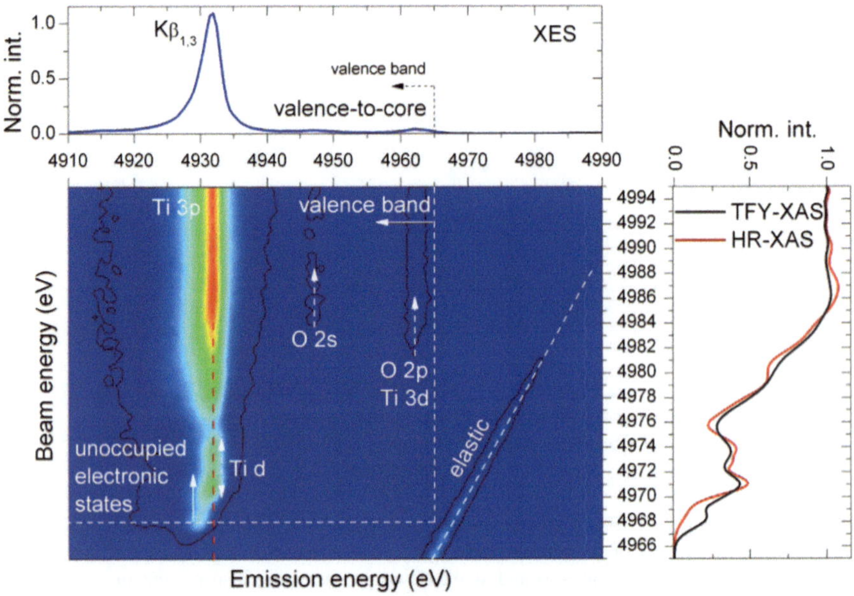

Fig. 4.5 The $K_{1,3}$ and valence-to-core RIXS for anatase. The electronic bands contributions are schematically marked in the figure. *Top* XES spectrum for incident X-ray energies tuned above the absorption edge. *Right* comparison of high-resolution XAS and total-fluorescence XAS spectra. Reproduced with permission from [41]

Fig. 4.6 (*Top*) Occupied (*blue*) and unoccupied (*red*) electronic states of titania extracted from the RIXS plane. The measured data are compared to DOS calculations (*bottom*) [41]. Reproduced with permission from [41]

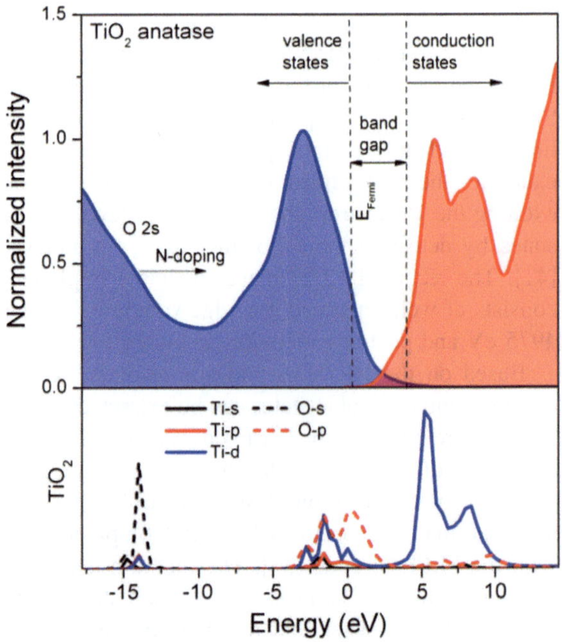

small. For the occupied electronic states just below the Fermi level an equal contribution of O-p and Ti-d orbitals is observed. An additional weak structure can be seen at around −15 eV that results from the 2s orbital of oxygen [41].

4.4 Examples

4.4.1 Carbon Monoxide Oxidation: Time-Resolved X-ray Absorption

The electronic structure of metal nanoparticles is strongly dependent on the presence and nature of adsorbates, and these adsorbates in turn affect the shape and intensity of features in the XAS spectrum. This makes XAS an attractive tool for investigating reactions like carbon monoxide oxidation, where multiple potential adsorbates are present and the catalyst structure depends on the reaction conditions. An early study showing the potential of the technique for investigating the adsorption of reactant molecules on a platinum catalyst was published by Safonova et al. in 2006 [7]. While this work did not constitute an *operando* experiment, it demonstrated the observation of differences in the electronic structure of platinum nanoparticles under realistic conditions. A particular emphasis was put on the comparison between conventional and HERFD XAS. While the former is able to distinguish between surface-oxidized, carbon-monoxide-covered, and bare platinum, the improved energy resolution of HERFD XAS gives additional electronic information. This information could be combined with theoretical calculations to show that atop adsorption of carbon monoxide is the dominant conformation under the measured conditions.

Based on this demonstration of the potential of XAS and HERFD XAS for studying platinum structure under reaction conditions, Singh et al. combined both techniques to investigate structural changes in platinum during light-off under realistic reaction conditions and with simultaneous monitoring of conversion [42]. The combination of both techniques is desirable, since the high time resolution of QEXAFS permits following the structural changes during light-off (Fig. 4.7), whereas the better energy resolution of HERFD XAS gives a more detailed picture of the initial and final states of the catalyst.

High-energy resolution off-resonant X-ray spectroscopy (HEROS) combines the attractive features of HERFD XAS and QEXAFS by enabling the recording of spectra with high time- and energy resolution. This was demonstrated by Szlachetko et al. for cyclic switching between oxygen and carbon monoxide over a platinum-based catalyst [43]. While the cyclical operation required for HEROS is distinct from the way most catalytic reactors are operated, the method gives fundamental insights into the steps required in the catalytic reaction under realistic conditions.

Oscillating reactions in carbon monoxide oxidation have also received considerable attention in XAS studies. The main motivation of such work was finding correlations between changes in structure and oxidation state of the metal in the

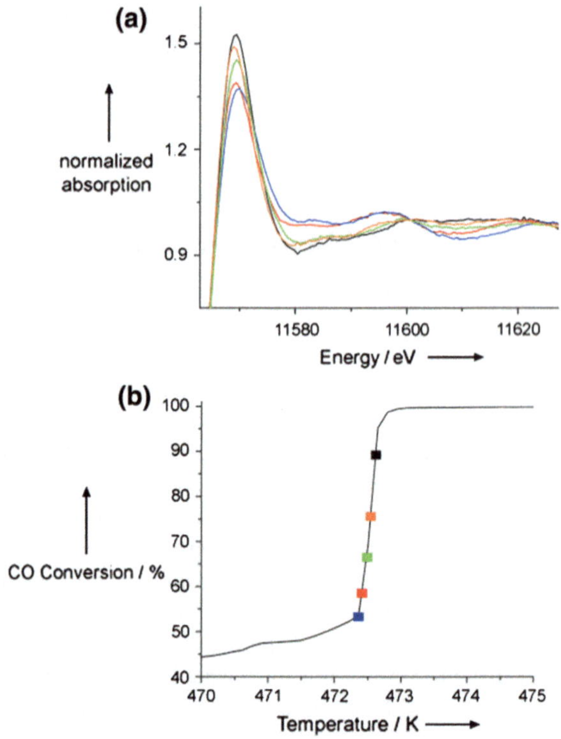

Fig. 4.7 XANES spectra (**a**) and carbon monoxide conversion (**b**) during heating-induced light-off in carbon monoxide oxidation over platinum. The Pt L_3 edge XANES spectra reveal that the transition from carbon-monoxide-covered platinum to oxidized platinum coincides with a sudden increase in activity

catalyst, and the oscillating conversion. Singh et al. used a combination of mass spectrometry, quick XAS and infrared spectroscopy to follow adsorbate concentrations, platinum oxidation state, and conversion during cooling, which includes significant oscillatory phenomena [44]. XANES was measured in multiple positions along the direction of flow, and it was found that further downstream in the reactor the catalyst reduced at lower conversion levels, i.e. that the reduction proceeded from the inlet side in a downstream direction.

Boubnov et al. used a combination of XAS and DRIFTS to investigate oscillatory phenomena in carbon monoxide oxidation during heating [45]. Comparing the changes in whiteline intensity with temperature at different positions, they observed that oscillations start at lower temperatures closer to the outlet side of the reactor. While this result may appear to contradict the results of Singh et al., it must be noted that one experiment was carried out while heating, and the other while cooling. Both experiments were carried out in excess oxygen, resulting in an increase of the oxygen/carbon monoxide ratio throughout the reactor during catalytic conversion. It is therefore expected that the reduction of the platinum during cooling initiates where conditions are most reducing (i.e. near the inlet), whereas oxidation during heating initiates where they are most oxidizing (i.e. near the outlet).

Figueroa and Newton used energy dispersive XAS and DRIFTS to investigate oscillatory behavior in carbon monoxide oxidation on a rhodium catalyst [46].

Unlike the examples discussed for platinum, the oscillations over the rhodium catalyst did not appear to be driven by changes in the oxidation state of the metal. Instead, they suggest a mechanism based on the presence of atomic carbon on the sample, which does not result in changes detectable by XAS.

4.4.2 Adsorption of Carbon Monoxide: High Energy Resolution XAS Spectra

Reactant adsorption is an essential step in every catalytic reaction, irrespective of the reaction mechanism. Carbon monoxide oxidation is generally assumed to occur by a Langmuir-Hinshelwood-type mechanism, in which both oxygen and carbon monoxide are chemisorbed on the surface and react with each other [47]. Carbon monoxide poisoning of a noble metal catalyst is often observed where the surface is covered by carbon monoxide. Desorption is needed to free a site for oxygen to react [48]. X-ray based methods provide fundamental insight into the bonding of carbon monoxide on nano-sized particles [49]. X-ray absorption spectra provide the element- and spin-selective density of states. The Pt L_3 edge probes the empty d density of states directly above the Fermi level and thus probes empty non-bonding and anti-bonding states. A disadvantage of adsorption edges of such high energy is that the spectra are broad and spectral detail may be lost. High energy resolution fluorescence detected (HERFD) XAS yields spectra of high energy resolution, which highlight many more spectral details. Such an experiment was pioneered on platinum by Sofanova et al., as discussed above [7]. The origin of spectral energy resolution is the lifetime of the final state; In a HERFD experiment, specific fluorescence decay channels are selectively probed. This fluorescence channel leaves a new core hole that has a lower core-hole lifetime broadening, resulting in less spectral broadening. It is essential to use a fluorescence/emission detector that has an energy resolution equal to or better than the lifetime broadening of the new final state. This results in spectra that have enhanced spectral resolution. Figure 4.8 shows typical Pt L_3 edge spectra of nano-sized platinum particles, here supported on alumina.

These spectra illustrate the empty d density of states of particles free of adsorbates, with carbon monoxide adsorbed, and after oxidation of their surface. Especially the spectrum of the particles covered in carbon monoxide is revealing. Compared to the spectrum of the particles free of adsorbates, an upwards energy shift is observed. Such shift in Fermi level identifies charge transfer from the platinum 5d orbitals to carbon monoxide. Moreover, the first intense spectral feature, the so-called whiteline, is split into a doublet, the high-energy component originating from a newly-formed anti-bonding orbital between platinum d and carbon and oxygen s and p orbitals. Carbon monoxide can adsorb on different sites on metal surfaces: on atop, bridged and threefold sites binding to respectively one, two, and three metal atoms. Because each such site has a different electronic configuration, XAS spectra are sensitive to the mode of bonding, which can be

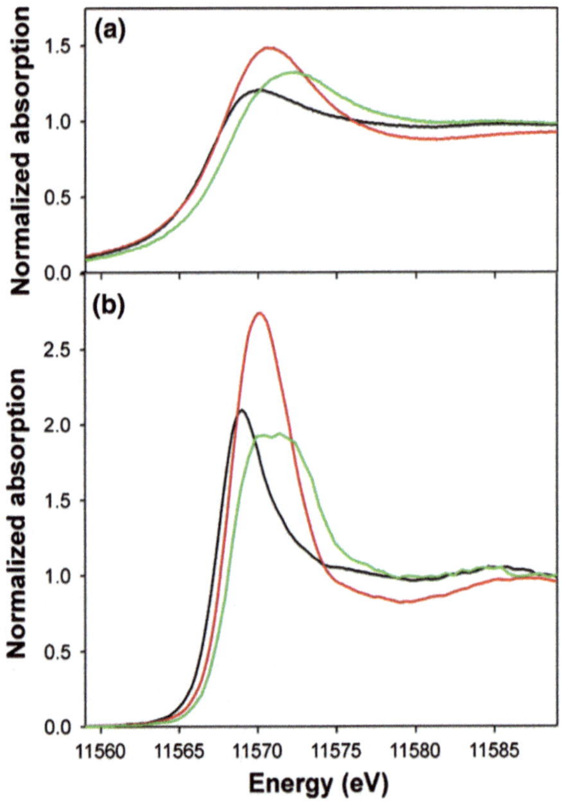

Fig. 4.8 Pt L_3 edge **a** XAS and **b** HERFD XAS of platinum nano-particles on alumina free of adsorbates (*black*), after surface oxidation (*red*), and covered in carbon monoxide (*green*). The higher energy resolution of HERFD XAS permits the differentiation between different possible carbon monoxide adsorption sites. Reproduced with permission from [7]

assessed by theoretical modeling relatively easily [4]. Fortuitously, the simulated Pt L_3 spectrum of atop bonded carbon monoxide deviates from other adsorption modes and can be quantitatively assessed. Theory thus unambiguously identifies the adsorption mode of carbon monoxide as bonded to a single metal atom in an atop fashion. In this case, the high energy resolution helped to unambiguously identify the adsorption site. Normal XAS showed the shift in absorption edge, however, did not resolve the doublet complicating unambiguous spectral assignment.

4.4.3 *Adsorption of Carbon Monoxide: Valence-to-Core RIXS*

Bonding originates from forming bonding and anti-bonding states [50]. In XAS, the orbitals above the Fermi level are probed, in XES those that are filled. Thus, an experiment, which probes the valence band directly probes the bonding states. Such an experiment has been done on the same platinum nano-particles [34] as the HERFD XAS study described above. In such a valence-to-core resonance inelastic

X-ray scattering (valence-to-core RIXS) experiment, an electron from a core level is excited (X-ray absorption) and the resulting fluorescence valence-to-core decay (X-ray emission) is detected as function of energy (Fig. 4.9). The resulting plot depicts the filled 5d density of states, the Fermi energy at 0 eV. The nanoparticles free of adsorbates have the d band cut through the Fermi level, whereas after carbon monoxide adsorption, a shift of the Fermi level is observed, which is illustrative of a band gap opening. The band shift is towards positive energy transfer, thus higher binding energy. Such a shift lowers the energy levels of the bonding and anti-bonding states with adsorbates, decreasing the reactivity [34] of the platinum nanoparticles. Thus, covering platinum particles with carbon monoxide not only occupies reactive sites, but also decreases the reactivity towards other reactants, such as oxygen. Overall, carbon monoxide passivates the nanoparticles (Fig. 4.9).

4.4.4 Time-Resolved X-ray Emission: Catalytically Active and Spectator Species

Heterogeneous catalysts generally consist of multiple components; Catalytic particles are attached to a support, which are often not inert. The support affects the catalytic performance by changing the electronic and/or geometric structure of the nanoparticles and they can even catalyze one or more reaction steps. Both factors play a role in oxidation of carbon monoxide over supported catalysts. Because of the poisoning effect of carbon monoxide on the reaction, as illustrated in Fig. 4.11, a support that helps to activate oxygen enhances the rate of reaction. Thus, redox-active supports yield more active catalysts. Ceria is a well-known redox-active material and is a component in many commercial catalysts, notably in automotive exhaust. Ceria is a so-called oxygen-storage material: Release of oxygen reduces Ce^{4+} to Ce^{3+}; Oxygen uptake re-oxidizes the Ce^{3+}. The oxygen-storage capacity is defined as the amount of oxygen that can be reversibly released under defined conditions. It relates to the fraction of Ce^{3+} that can reversibly be formed. The oxygen in the support can thus be used to maintain an oxidative reaction, whenever gas-phase oxygen becomes depleted. It is generally assumed that both the oxygen-storage capacity and Ce^{3+} play essential roles in the catalytic mechanism, however, their precise contributions are debated. A recent study, employing transient conditions, identified the role of Ce^{3+} in the oxidation of carbon monoxide. The study emphasized the existence of spectator species that do not contribute to catalytic conversion [14].

Figure 4.10a shows schematically the experimental setup that was used to measure time-resolved resonance X-ray emission spectrum (RXES) of nano-sized platinum particles on ceria. The catalyst was positioned within a capillary reactor, which was connected at one side to a gas-dosing system and at the other to a mass spectrometer. The gas-dosing system allowed precise control of the gas composition, which was periodically changed from containing oxygen and carbon monoxide in a ratio of four

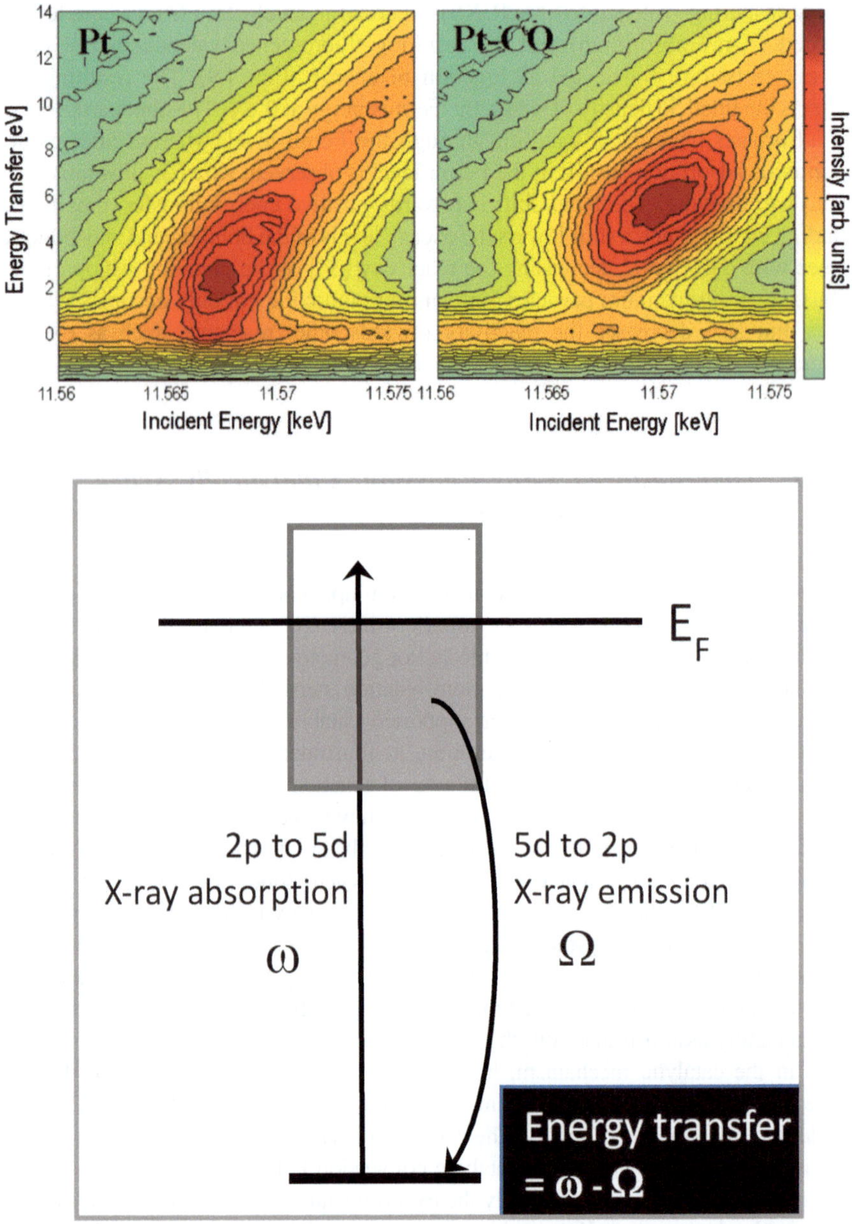

Fig. 4.9 (*Top*) $2p_{3/2}$ valence-to-core RIXS of alumina-supported platinum nanoparticles before (*left*) and after (*right*) adsorption of carbon monoxide. The energy transfer is depicted as function of excitation or absorption energy. (*bottom*) Energy transition in the RIXS experiment. The incident energy excites an electron from the 2p shell into the empty 5d band; the emission from the filled 5d band into the resulting core hole is detected with high energy resolution. The relative energies are not drawn to scale. Top figure reproduced with permission from [34]

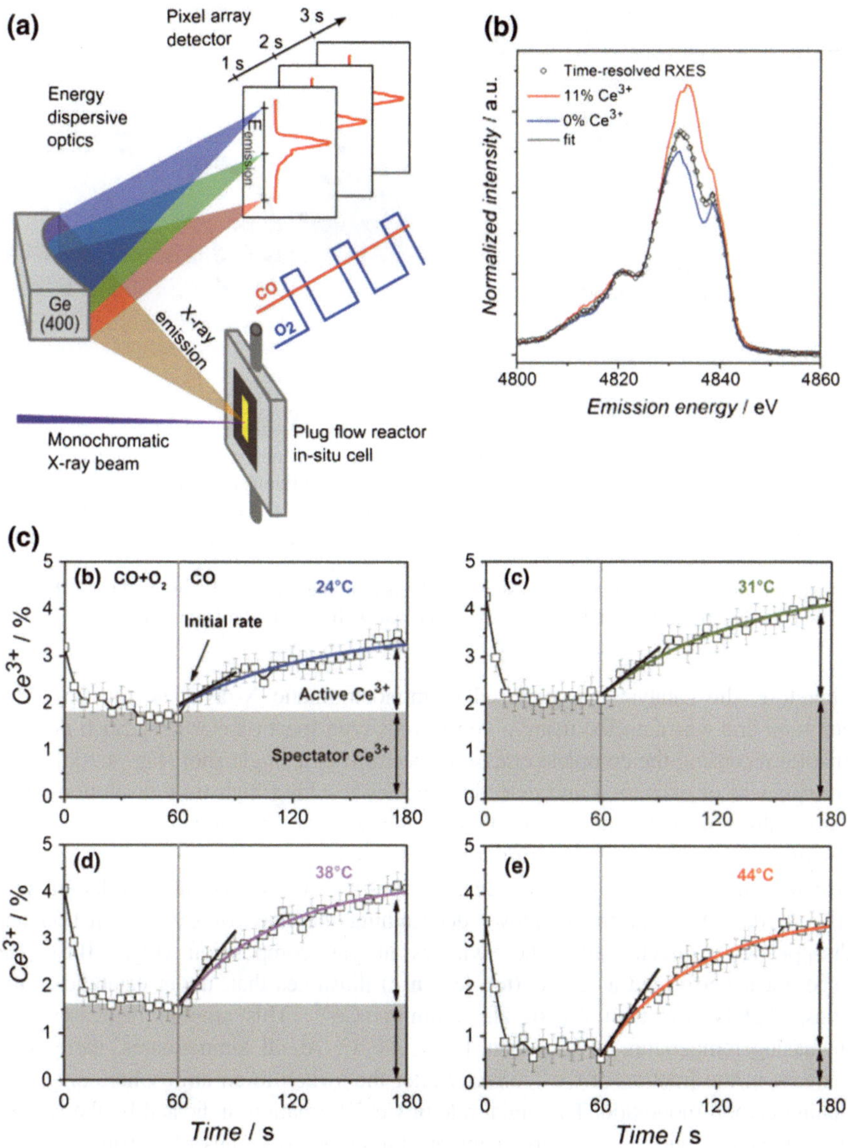

Fig. 4.10 **a** Setup to measure time-resolved RXES spectra in a single shot. **b** RXES spectra of materials with known amounts of Ce^{3+} and Ce^{4+} used to quantify the spectra during the transient experiment. **c** Development of the amounts of Ce^{3+} in Pt/ceria after the switch from carbon monoxide to oxygen/carbon monoxide at four different temperatures. Reproduced with permission from [14]

to one to containing only carbon monoxide. The conditions were chosen such that the maximum conversion always remained below 10% to ensure differential conditions. The rate of structural change of cerium was followed by time-resolved RXES.

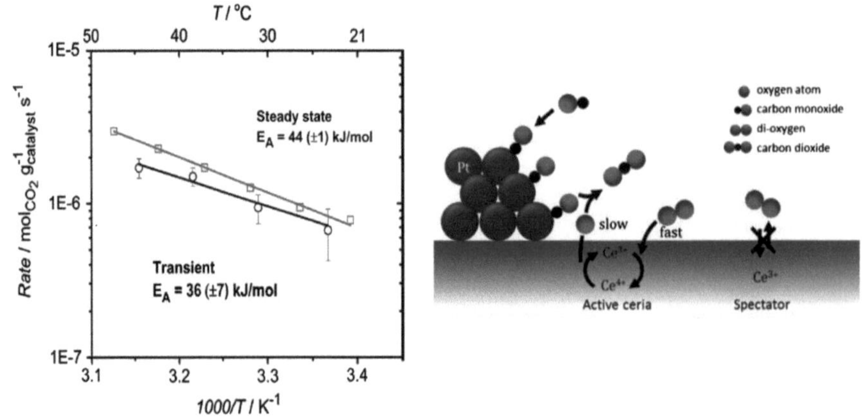

Fig. 4.11 (*Left*) Arrhenius plot of rate of steady-state carbon monoxide oxidation (*red*) and of initial rate of Ce^{4+} reduction after the switch from catalytic conditions to carbon monoxide. (*Right*) Possible reaction mechanism over ceria supported platinum. The platinum nanoparticles capture carbon monoxide, which reacts at the metal-support interface with an oxygen atom from the support, thereby reducing Ce^{4+} to Ce^{3+}. If oxygen is present in the gas phase, Ce^{3+} is very rapidly re-oxidized, making the active Ce^{3+} a short-lived reaction intermediate. Long-lived Ce^{3+} does not participate directly in the catalytic cycle. Reproduced with permission from [14]

Therefore, the catalyst was exposed to monochromatic X-rays and the $2p_{3/2}5d_{5/2}$ emission line was detected using a von Hamos-type spectrometer [51]. Such detector enables recording the complete emission spectrum in a single shot (Fig. 4.10a). Such arrangement of excitation and detection enables reaching high time resolutions, fundamentally limited by the readout time of the detector and the signal-to-noise ratio. Continuous measuring during multiple switches increases the signal-to-noise ratio. Because the $2p_{3/2}5d_{5/2}$ emission line is sensitive to the cerium oxidation state (Fig. 4.10b), the experiment allows determining the rates of Ce^{3+} formation and disappearance directly after the switches in gas composition (Fig. 4.10c). The experiment performed at 24 °C (top left in c) illustrates that, under catalytic conditions, slightly less than 2% of all cerium is Ce^{3+}. This amount decreases with increasing temperature to less than 1% at 44 °C. At all temperatures, there is an increase in the fraction of Ce^{3+} directly after the switch to an atmosphere only containing carbon monoxide. The initial rate of Ce^{3+} formation, indicated by the straight lines, increases with temperature enabling plotting an Arrhenius plot. This plot, after correction for stoichiometry, is compared to the rate of carbon monoxide oxidation under steady state conditions (Fig. 4.11 left), which was measured under identical conditions in the same reactor using gas chromatography. The lines measured by RXES and kinetic measurements coincide and are within the accuracy of measurements the same. This illustrates that directly after the switch, the reaction continues with exactly the same rate and that the oxygen atom originates from ceria. The consequence is that reduction of ceria is coupled to the rate-limiting step. Re-oxidation that occurred after the switch back to carbon monoxide and oxygen was very rapid and too fast to determine its kinetics. Thus, under steady-state catalytic conditions, the catalytically active cerium is virtually completely Ce^{4+}, and Ce^{3+} is a short-lived reaction intermediate (Fig. 4.11 right). Any Ce^{3+} that is present under steady-state conditions, is a spectator species and does not directly play a role in the reaction mechanism.

4.5 Conclusions

XAS and XES are excellent methods to identify the structure of catalysts under reaction conditions and within an actual reactor. They enable to determine the local geometric and electronic structure of a catalyst element specifically and with high time- and space resolution. Because the structure of platinum catalysts for the oxidation of carbon monoxide varies strongly with conditions, structural changes within a single reactor may occur. The different phases have different activity and, therefore, the precise relationship between structure of active sites and activity has in most cases not been established. The structure of the phase that is responsible for activity and that of the active site that performs the catalytic turnover must be distinguished.

Transient, modulated, and space-resolved experiments have established the structure of the phase that is associated with the low-activity phase; The platinum surface is maximally covered with carbon monoxide and oxygen activation over the metal only occurs after partial carbon monoxide desorption, which frees sites for oxygen activation. Much higher rates occur in case oxygen is activated on the support, such as ceria. Reaction than occurs at the metal—support interface.

The structure of platinum in the highly active state is much less established. It is associated with oxygen at the surface. The catalyst activity has been shown to scale with the presence of a platinum surface oxide, however, also reduced states are suggested. The exact structure of the surface oxide is not established, other than that it is disordered and defective. It can be speculated that such sites are responsible for binding carbon monoxide and that oxygen atoms in its vicinity react to form carbon dioxide.

It is safe to say that the site on a platinum catalyst that is responsible for oxygen activation and its subsequent reaction to carbon monoxide has not been unambiguously established. It is very well possible that within a single reactor, multiple structures are responsible for oxygen activation. Part of the origin of the absence of agreement originates from often strongly varying conditions, which affect catalyst structure and thus influence the outcome of the experiment. A complete description of active sites can only be achieved after quantitative determination of local gas composition and catalyst structure throughout the whole catalytic reactor. Combining such an experiment with complementary methods, such as vibrational spectroscopy enhances the chances of success.

References

1. S. Bordiga et al., Reactivity of surface species in heterogeneous catalysts probed by in situ X-ray absorption techniques. Chem. Rev. **113**(3), 1736–1850 (2013)
2. C. Lamberti et al., XAS Spectroscopy: related techniques and combination with other spectroscopic and scattering methods, in X-Ray Absorption and X-Ray Emission Spectroscopy: Theory and Applications, ed. by J.A. van Bokhoven, C. Lamberti (Wiley, Chichester (UK), 2016), pp. 303–350
3. P.M. Abdala et al., Scientific opportunities for heterogeneous catalysis research at the SuperXAS and SNBL beam lines. CHIMIA Int. J. Chem. **66**(9), 699–705 (2012)
4. D.E. Ramaker, D.C. Koningsberger, The atomic AXAFS and [capital Delta][small mu] XANES techniques as applied to heterogeneous catalysis and electrocatalysis. Phys. Chem. Chem. Phys. **12**(21), 5514–5534 (2010)
5. D.C. Koningsberger et al., XAFS spectroscopy; fundamental principles and data analysis. Top. Catal. **10**(3–4), 143–155 (2000)
6. P. Glatzel, U. Bergmann, High resolution 1s core hole X-ray spectroscopy in 3d transition metal complexes—electronic and structural information. Coord. Chem. Rev. **249**(1–2), 65–95 (2005)
7. O.V. Safonova et al., Identification of CO adsorption sites in supported Pt catalysts using high-energy-resolution fluorescence detection X-ray spectroscopy. J. Phys. Chem. B **110**(33), 16162–16164 (2006)
8. J. Singh, C. Lamberti, J.A. van Bokhoven, Advanced X-ray absorption and emission spectroscopy: in situ catalytic studies. Chem. Soc. Rev. **39**(12), 4754–4766 (2010)
9. F. Bonino et al., Catalyst characterization by Xas And Xes spectroscopies: in situ and operando experiments, in *Synchrotron Radiation*, ed. by S. Mobilio, F. Boscherini, C. Meneghini (Springer, Berlin, 2015), pp. 717–736
10. M. Dreher et al., Design of a continuous-flow reactor for in situ x-ray absorption spectroscopy of solids in supercritical fluids. Rev. Sci. Instr. **83**(5), 054101 (2012)
11. D. Ferri, M.A. Newton, M. Nachtegaal, Modulation excitation X-ray absorption spectroscopy to probe surface species on heterogeneous catalysts. Top. Catal. **54**(16), 1070–1078 (2011)
12. C.F.J. König et al., Quantitative analysis of modulated excitation X-ray absorption spectra: enhanced precision of EXAFS fitting. J. Phys. Chem. C **116**(37), 19857–19866 (2012)
13. D. Ferri et al., Synchrotron high energy X-ray methods coupled to phase sensitive analysis to characterize aging of solid catalysts with enhanced sensitivity. Phys. Chem. Chem. Phys. **15** (22), 8629–8639 (2013)
14. R. Kopelent et al., Catalytically active and spectator Ce^{3+} in ceria-supported metal catalysts. Angew. Chem. Int. Ed. **54**(30), 8728–8731 (2015)
15. R. Frahm, J. Stötzel, D. Lützenkirchen-Hecht, advancing time-resolved methods in monitoring and characterization of catalysts. Synchrotron Radiat. News **22**(2), 6–11 (2009)
16. M. Nachtegaal et al., QEXAFS: techniques and scientific applications for time-resolved XAS, in *X-Ray Absorption and X-Ray Emission Spectroscopy: Theory and Applications*, ed. by J.A. van Bokhoven, C. Lamberti (Wiley, Chichester (UK), 2016), pp. 155–183
17. M.A. Newton, A.J. Dent, J. Evans, Bringing time resolution to EXAFS: recent developments and application to chemical systems. Chem. Soc. Rev. **31**(2), 83–95 (2002)
18. O. Mathon, I. Kantor, S. Pascarelli, Time-resolved XAS using an energy dispersive spectrometer: techniques and applications, in *X-Ray Absorption and X-Ray Emission Spectroscopy: Theory and Applications*, ed. by J.A. van Bokhoven, C. Lamberti (Wiley, Chichester (UK), 2016), pp. 185–212
19. J.J. Bravo-Suárez et al., Transient technique for identification of true reaction intermediates: hydroperoxide species in propylene epoxidation on gold/titanosilicate catalysts by X-ray absorption fine structure spectroscopy. J. Phys. Chem. C **112**(4), 1115–1123 (2008)
20. D. Baurecht, U.P. Fringeli, Quantitative modulated excitation Fourier transform infrared spectroscopy. Rev. Sci. Instr. **72**(10), 3782–3792 (2001)

21. A. Eyssler et al., Improvement of catalytic activity of LaFe0.95Pd0.05O3 for methane oxidation under transient conditions. J. Phys. Chem. C **115**(4), 1231–1239 (2011)
22. D. Ferri et al., First steps in combining modulation excitation spectroscopy with synchronous dispersive EXAFS/DRIFTS/mass spectrometry for in situ time resolved study of heterogeneous catalysts. Phys. Chem. Chem. Phys. **12**(21), 5634–5646 (2010)
23. E.E. Ortelli, A. Wokaun, Use of periodic variations of reactant concentrations in time resolved FTIR studies of heterogeneously catalysed reactions. Vib. Spectrosc. **19**(2), 451–459 (1999)
24. T. Bürgi, A. Baiker, In situ infrared spectroscopy of catalytic solid–liquid interfaces using phase-sensitive detection: enantioselective hydrogenation of a pyrone over Pd/TiO2. J. Phys. Chem. B **106**(41), 10649–10658 (2002)
25. A. Urakawa et al., Combined, modulation enhanced X-ray powder diffraction and raman spectroscopic study of structural transitions in the spin crossover material [Fe(Htrz)2(trz)](BF4). J. Phys. Chem. C **115**(4), 1323–1329 (2011)
26. D. Chernyshov et al., Kinematic diffraction on a structure with periodically varying scattering function. Acta Crystallographica Section A **67**(4), 327–335 (2011)
27. A. Rochet et al., Influence of the preparation conditions of oxidic NiMo/Al2O3 catalysts on the sulfidation ability: a quick-XAS and raman spectroscopic study. J. Phys. Chem. C **119** (42), 23928–23942 (2015)
28. W.H. Cassinelli et al., Multivariate curve resolution analysis applied to time-resolved synchrotron X-ray absorption spectroscopy monitoring of the activation of copper alumina catalyst. Catal. Today **229**, 114–122 (2014)
29. A. Kotani, S. Shin, Resonant inelastic X-ray scattering spectra for electrons in solids. Rev. Mod. Phys. **73**(1), 203–246 (2001)
30. U. Bergmann et al., High-resolution X-ray spectroscopy of rare events: a different look at local structure and chemistry. J. Synchrotron Radiat. **8**(2), 199–203 (2001)
31. J.L. Campbell, T. Papp, Widths of the atomic K-N7 levels. At. Data Nucl. Data Tables **77**(1), 1–56 (2001)
32. J.J. Rehr et al., Parameter-free calculations of X-ray spectra with FEFF9. Phys. Chem. Chem. Phys. **12**(21), 5503–5513 (2010)
33. J.J. Rehr et al., Ab initio theory and calculations of X-ray spectra. C.R. Phys. **10**(6), 548–559 (2009)
34. P. Glatzel et al., In situ characterization of the 5d density of states of Pt nanoparticles upon adsorption of CO. J. Am. Chem. Soc. **132**(8), 2555–2557 (2010)
35. E. Kleymenov et al., Five-element Johann-type X-ray emission spectrometer with a single-photon-counting pixel detector. Rev. Sci. Instrum. **82**(6), 065107 (2011)
36. E.M. Alayon et al., Catalytic conversion of methane to methanol over Cu-mordenite. Chem. Commun. **48**(3), 404–406 (2012)
37. O.V. Safonova et al., Local environment of vanadium in V/Al/O-mixed oxide catalyst for propane ammoxidation: Characterization by in situ valence-to-core X-ray emission spectroscopy and X-ray absorption spectroscopy. J. Catal. **268**(1), 156–164 (2009)
38. G. Smolentsev et al., X-ray emission spectroscopy to study ligand valence orbitals in Mn coordination complexes. J. Am. Chem. Soc. **131**(36), 13161–13167 (2009)
39. U. Bergmann et al., Chemical dependence of interatomic X-ray transition energies and intensities—a study of Mn $K\beta''$ and $K\beta 2$, 5 spectra. Chem. Phys. Lett. **302**(1–2), 119–124 (1999)
40. C.J. Pollock, S. DeBeer, Valence-to-core X-ray emission spectroscopy: a sensitive probe of the nature of a bound ligand. J. Am. Chem. Soc. **133**(14), 5594–5601 (2011)
41. J. Szlachetko, J. Sa, Rational design of oxynitride materials: from theory to experiment. CrystEngComm **15**(14), 2583–2587 (2013)
42. J. Singh et al., Generating highly active partially oxidized platinum during oxidation of carbon monoxide over Pt/Al2O3. In situ, time-resolved, and high-energy-resolution X-ray absorption spectroscopy. Angew. Chem. Int. Ed. **47**(48), 9260–9264 (2008)

43. J. Szlachetko et al., Subsecond and in situ chemical speciation of Pt/Al2O3 during oxidation-reduction cycles monitored by high-energy resolution off-resonant X-ray spectroscopy. J. Am. Chem. Soc. **135**(51), 19071–19074 (2013)
44. J. Singh et al., Dynamic structure changes of a heterogeneous catalyst within a reactor: oscillations in CO oxidation over a supported platinum catalyst. Chemcatchem **2**(6), 653–657 (2010)
45. A. Boubnov et al., Oscillatory CO oxidation over Pt/Al2O3 catalysts studied by in situ XAS and drifts. Top. Catal. **56**(1–8), 333–338 (2013)
46. S.J.A. Figueroa, M.A. Newton, What drives spontaneous oscillations during CO oxidation using O2 over supported Rh/Al2O3 catalysts? J. Catal. **312**, 69–77 (2014)
47. M. Chen, Y. Zheng, H. Wan, Kinetics and active surfaces for CO oxidation on Pt-group metals under oxygen rich conditions. Top. Catal. **56**(15–17), 1299–1313 (2013)
48. X. Su et al., High-pressure CO oxidation on Pt(111) monitored with infrared—visible sum frequency generation (SFG). J. Am. Chem. Soc. **119**(17), 3994–4000 (1997)
49. N. Guo et al., Determination of CO, H2O and H2 coverage by XANES and EXAFS on Pt and Au during water gas shift reaction. Phys. Chem. Chem. Phys. **12**(21), 5678–5693 (2010)
50. B. Hammer, J.K. Norskov, Why gold is the noblest of all the metals. Nature **376**(6537), 238–240 (1995)
51. J. Szlachetko et al., A von Hamos x-ray spectrometer based on a segmented-type diffraction crystal for single-shot x-ray emission spectroscopy and time-resolved resonant inelastic x-ray scattering studies. Rev. Sci. Instr. **83**(10), 103105 (2012)
52. J. Szlachetko et al., In situ hard X-ray quick RIXS to probe dynamic changes in the electronic structure of functional materials. J. Electron Spectrosc. Relat. Phenom. **188**, 161–165 (2013)

Chapter 5
Development of *Operando* Transmission Electron Microscopy

Patricia Jane Kooyman

Abstract Traditional (high-resolution) Transmission Electron Microscopy (TEM) is limited to high vacuum environments due to interaction of the electron beam with gases, leading to noise and decreasing the resolution. This means only static materials can be studied, far away from realistic reaction conditions in gas atmosphere and at elevated temperatures. This chapter describes the development of equipment for *operando* TEM and the first studies of real in situ catalyst preparation and catalysis.

5.1 Introduction

Since its development by Knoll and Ruska [1], the trend in transmission electron microscopy (TEM) has been towards better vacuum. Lower pressure decreases the amount of random collisions of the electrons with gas molecules and thus decreases noise and improves resolution. However, this means we are always studying static inorganic materials—close to room temperature and away from reacting gases. Especially for heterogeneous catalysis research this a tremendous disadvantage, as we are fairly sure that the structure of the catalyst changes with adsorbed species and thus also during the catalytic reaction. These changes cannot be studied with the 'before-and-after' approach of standard ex situ TEM. Although protective atmosphere transfer holders are available to perform quasi in situ TEM and prevent exposure to air and moisture during sample preparation and transport to the microscope, these still force us to study static catalysts at 'frozen' vacuum and room temperature conditions.

Over the past decades, scientists have been devising ways to insert gases into TEMs without destroying either the resolution of the microscope or—even worse—the electron source. Due to the nature of the process of emitting high-energy electrons, the electron sources of TEMs are very sensitive to any gases present.

P.J. Kooyman (✉)
Chem Eng, University of Cape Town, Private Bag X3, 7700 Rondebosch, South Africa
e-mail: patricia.kooyman@uct.ac.za

J. Frenken and I. Groot (eds.), Operando *Research in Heterogeneous Catalysis*,
Springer Series in Chemical Physics 114, DOI 10.1007/978-3-319-44439-0_5

They will react even with traces of so-called 'inert' gases such as noble gases or nitrogen gas and be destroyed in the process.

In order to study catalytic processes in situ or *operando* (live as they are occurring), the samples also need to be heated, up to about 600 °C depending on the exact reaction under study. Heating holders were already available for standard vacuum-mode TEM, but the heaters need to be integrated into the holders or devices used in the in situ mode. A recent review [2] also considers the windowed-cell systems we are considering here, but covers a very broad range of applications (even including biological applications) and thus does not include the thorough review of catalytic applications given here.

The first successes with gases present inside the TEM during the imaging process led to the development of the now commercially available [3] Environmental TEM (ETEM), where localised gas pressures up to 50 mbar do not destroy the electron source but gas pressures above ~5 mbar do destroy the resolution.

Recent efforts have led to windowed cell systems, which have culminated in a nanoreactor system allowing pressures up to several bars. This chapter will describe the development of gas-phase in situ TEM from its roots at very low pressures until the highest currently available pressures. The development will be illustrated using examples from catalysis research.

Please note that in literature 'in situ TEM' often refers to performing certain processes in the vacuum of a traditional TEM, thus without the presence of gases or liquids. Examples of processes studied in this way are heating, electromigration and applying mechanical stress [4].

5.2 Aperture-Based Systems

The first systems addressing adaptations to TEM in order to be able to admit gases while simultaneously imaging were based on adding several additional apertures to existing TEM designs. The system with which to date most results have been obtained is the ETEM (Environmental TEM). This system is also referred to as CAEM (Controlled Atmosphere Electron Microscopy), depending on who developed it. However, as the only system commercially available at the time of writing is called ETEM [3], we will use that term for this chapter.

The ETEM requires modifying a standard TEM by incorporation of an ECELL (environmental cell, see Fig. 5.1 and [5]). In practice this means acquisition of a dedicated and expensive TEM. Implementing an ECELL into a TEM consists of adapting the pole pieces and the column, and adding some more apertures next to the ones that are already present in a standard TEM. These adaptations allow gases to be inserted around the sample (through the pole pieces) and to be pumped out before they reach the electron source (through the pole pieces and the additional pumping lines installed at the height of condenser and selected area apertures). The additional cell apertures prevent most of the gas from diffusing towards the electron source and restrain the path length of the electrons through the highest gas density.

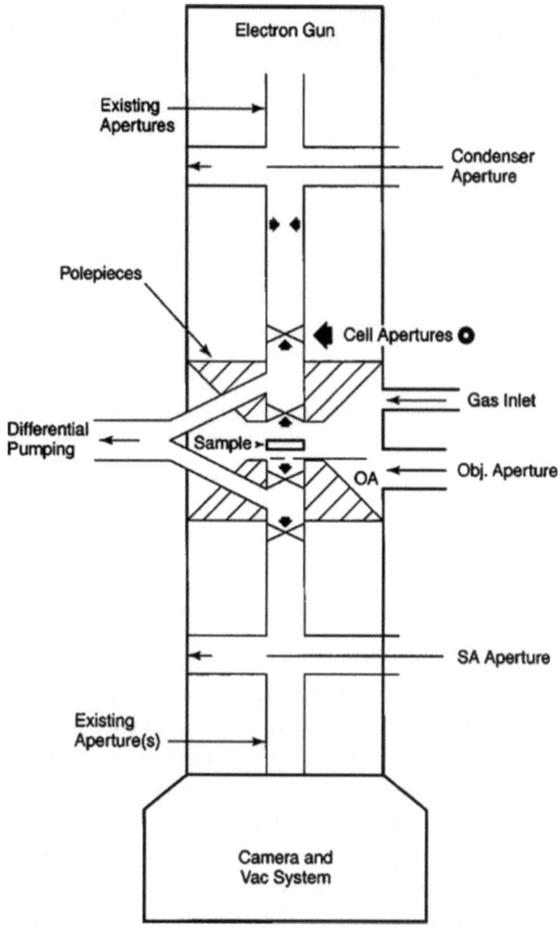

Fig. 5.1 Schematic cross-section of the ETEM. From [5]

Pioneering work in the development and application of ETEM was done by Hashimoto et al. [6]. Their system is already the basic design of an ECELL within a TEM. It allows pressures up to 400 mbar and temperatures up to 1000 °C, 'although in practice one tends to work at much lower pressures' according to Baker [7], who applied the system to many different catalytic studies. Baker already recognised the possible effects of electron beam interaction both on the catalyst and on the gas phase, and the important loss of resolution. This set-up required tedious sample preparation, as the material to be studied had to be present at the exact location indicated by the arrow labelled 'specimen' in Fig. 5.2. Thus, natural single-crystal graphite first had to be cleaved, cleaned, and mounted on glass slides. Then, successive layers had to be removed using Scotch tape until optically transparent areas remained attached to the glass slides. These optically transparent graphite crystals are often also transparent to electrons. They had to be released onto a clean water bath, picked up on a sapphire plate and degassed for half an hour

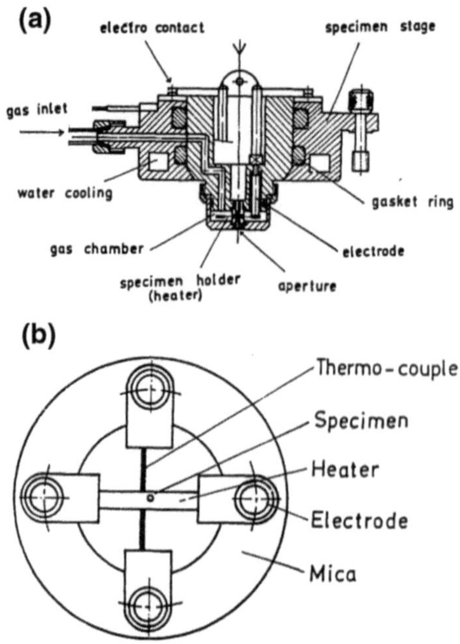

Fig. 5.2 Sample holder for CAEM. **a** Cross-section; **b** *top view* of specimen holder (heater). From [6]

at 900 °C before being floated on water again to be picked up on the specimen holder of Fig. 5.2 and dried. The model catalyst support (e.g. silica) then had to be prepared on the graphite by vapour deposition or sputtering. Electron—transparent alumina can also be prepared from aluminium foil by another tedious procedure. Finally, the metal catalyst had to be prepared on the model support material using evaporation from high-purity metal wires or by spraying a soluble salt solution.

Using this set-up, Baker became famous for studying the catalytic growth of carbon nanofibers [8]. At the time his papers were published it was not customary to also publish movies, but more recent movies showing the same formation of carbon nanofibers can be found on-line [9].

The main drawback of the ETEM is that only low pressures can be used (up to about 50 mbar without destroying the electron source if that is a FEG (Field Emission Gun), but only up to about 5 mbar without destroying the resolution). This was already illustrated by Hashimoto et al. [6] and is shown here in Fig. 5.3. The same area of zinc oxide nanocrystals is shown in an atmosphere of air inside the TEM. At 10^{-4} mbar (top image) the crystals are imaged with clear edges. At 133 mbar (central image) the image is already blurred, and at 400 mbar (bottom image) the crystals have become vague. This is not due to reaction taking place (the images were taken in air at room temperature, where zinc oxide is stable), but purely to scattering of the electron beam by the gas molecules present in their path.

Although the commercially available ETEM uses fairly standard heating holders and thus specimens can be applied to fairly standard TEM grids (though copper grids will be too unstable for most applications; platinum and gold grids are usually

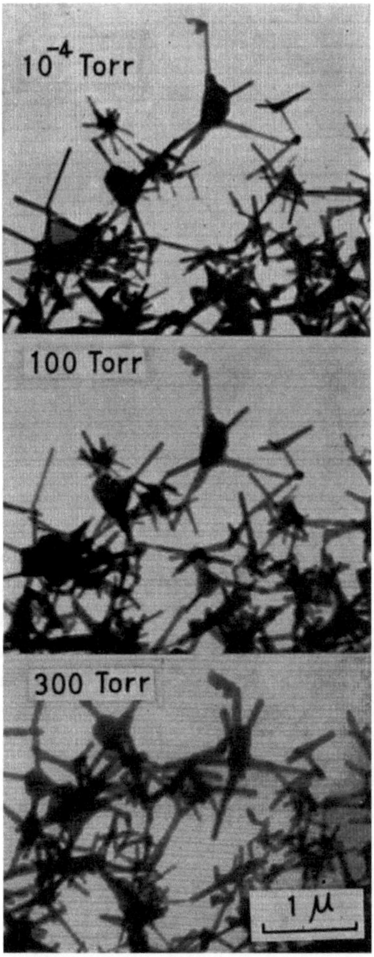

Fig. 5.3 Images of the same area of zinc oxide nanocrystals in air. *Top* 10^{-4} mbar; *middle* 133 mbar; *bottom* 400 mbar. From [6]

a better choice), another drawback is that the whole sample holder is exposed to the gases and can thus react with the gases. An example of what happened to a heating holder in CO gas at 40 mbar and 400 °C inside an ETEM is given in Fig. 5.4. After studying the carburisation of iron oxide [10], the whole Pt heater cup was covered in a thick layer of carbon deposit. This does not necessarily hamper the study of the processes at the catalyst itself, but it will hamper study of the gas composition during the process of interest. It might also give rise to a different gas composition at the catalyst than what is fed into the ETEM. An excellent review of the application of ETEM in catalysis research was published recently, combined with a review of high pressure STM work in the same field [11]. A whole book dedicated to the development and applications of ETEM was also published recently, with Chap. 8 dedicated to catalysis research [12].

Fig. 5.4 Sample holder heater cup after performing an experiment in 40 mbar CO gas and heating to 400 °C in an ETEM. Part of the carbon deposit (*dark grey* powder) has been scraped off (metal-coloured area) to show the extent of the coking

5.3 Windowed Cell Systems

In order to protect the electron source from gases and simultaneously reduce the electron path length through the gas layer by containing the gases within a limited volume, windowed cell systems have been developed.

JEOL offers a windowed cell system built into a holder on demand (their website does not mention it) [13]. It has been used by a handful of people, notably Giorgio [14, 15]. Figure 5.5 shows the design of this system, which can still only be operated up to 350 °C and 10 mbar, or up to 30 mbar at room temperature. But sample loading is relatively easy as a powder can be directly applied to the heating wire for the former set of conditions or to the bottom windows for the latter set of conditions.

With the development of MEMS (Micro-Electro-Mechanical-Systems) engineering, both the windows and the heater systems could be miniaturised. Smaller windows and controlled morphological design of the windows mean larger pressure

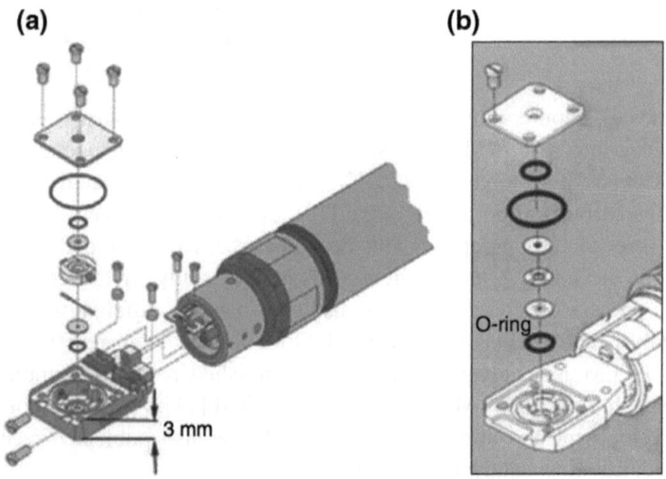

Fig. 5.5 The JEOL ECELL. **a** For heating. **b** For room temperature. From [15]

differences between the confined gas phase and the vacuum of the TEM become possible. Smaller and better controlled heating systems mean localised heating and thus the desired reaction with the gas(es) occurring only at the area under investigation, with minimalised temperature-induced drift as a huge bonus. These MEMS-based systems consist of chip-based sample holders that are placed into adapted TEM holders, and can be used with regular TEMs of several brands, avoiding adaptation of the microscopes. Although not commercially available, the best performance to date has been achieved with the nanoreactor system developed at Delft University of Technology (first published in 2008 [16]). This will be discussed in more detail below.

Three MEMS-based systems are available commercially. The first publications concerning the Protochips Atmosphere system [17] date from 2010 [18], but the first applications were published in 2012 [19, 20]. The first publications concerning the Hummingbird Scientific Gas Flow system [21] date from 2012 [22], and application publications followed already in 2013 [23, 24]. As of the time of writing, no scientific publications had appeared yet using the DENSsolutions system [25]. Like the system Giorgio used, these commercially available systems are O-ring-based systems. These systems have the advantage of easy loading of powdered samples, but the disadvantage of large dead volumes of gas. This means that monitoring changes in the solid-state material and in the gas phase while changing the inlet gas composition is hampered by slow changes in gas composition at the observation area. Figure 5.6 shows the Protochips design, Fig. 5.7 the Hummingbird design, and Fig. 5.8 the DENSsolutions design.

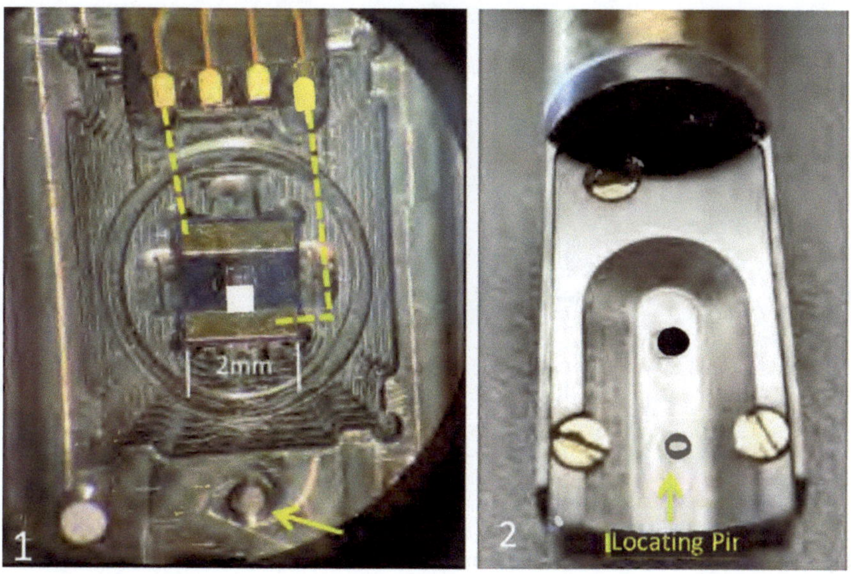

Fig. 5.6 The Protochips design. **1** Interior view of the gas-cell holder showing the heater, electrical contacts and locating pin (*arrow*) for the retainer plate. **2** Tip of the gas-cell holder showing the retaining plate for the assembled cell and the locating pin (*arrow*). From [20]

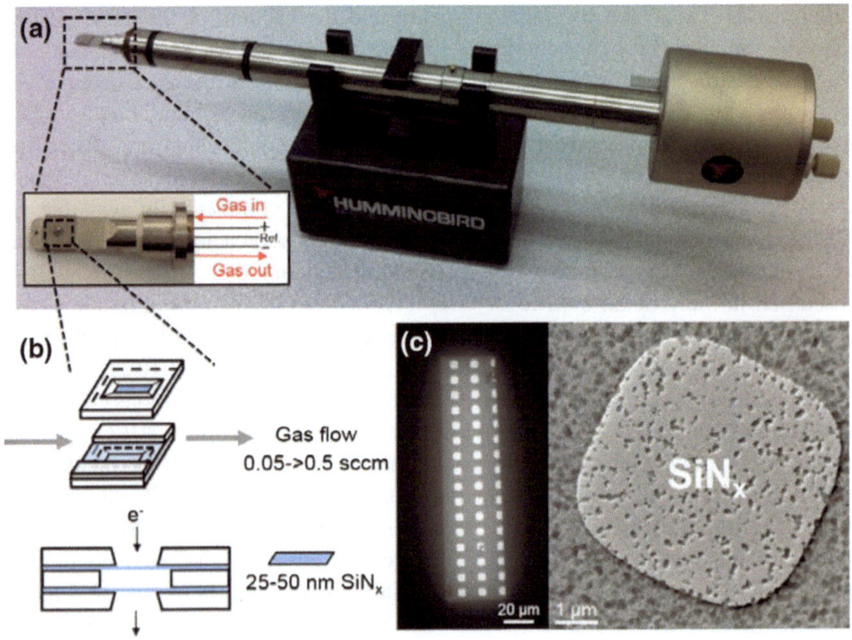

Fig. 5.7 The Hummingbird design. **a** Gas flow heating holder. **b** MEMS-type gas flow cell. **c** The electron transparent windows. From [23]

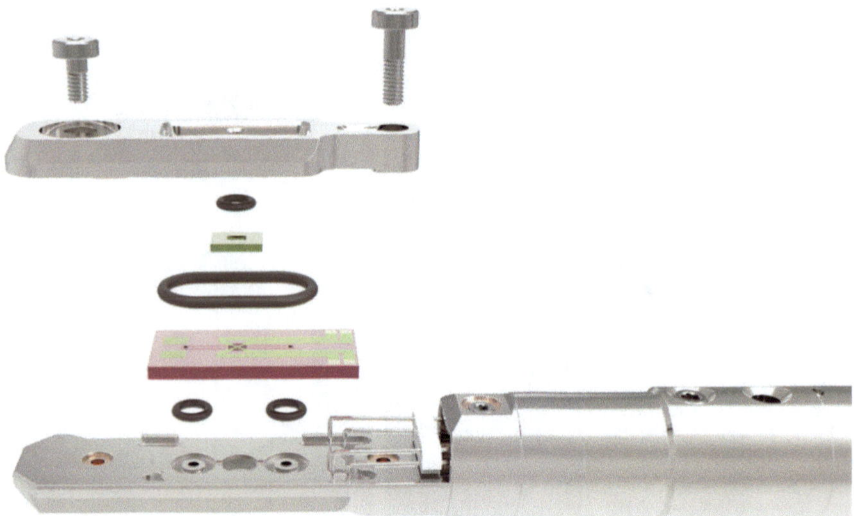

Fig. 5.8 The DENSsolutions design. From [25]

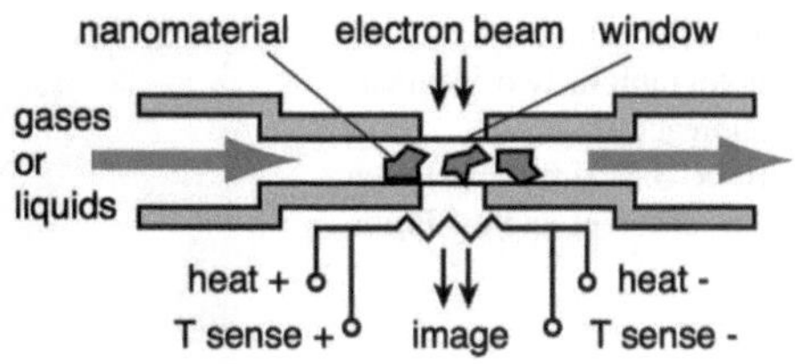

Fig. 5.9 Schematic cross-section of a nanoreactor. From [30]

The MEMS-based nanoreactor system has been developed by a large team based at Delft University of Technology but also involving Leiden University and several companies [26]. It is based on a windowed MEMS chip, where the gas channel is narrow compared to the width of the chip [16, 27, 28]. This means all gas is forced over the heater area and thus dead volume within the nanoreactor itself is practically non-existent. Dead volume in the whole system is determined by the gas-dosing system and the length and diameter of the gas lines used. The basic design consists of two dies that have to be aligned in order to have the electron-transparent windows overlap perfectly. Chapter 6 of [12] offers a very detailed description of this system. A schematic cross-section of a nanoreactor is shown in Fig. 5.9 and a schematic drawing of a complete nanoreactor in Fig. 5.10.

Figure 5.11 shows an optical photograph of two identical nanoreactors seen from different angles, and Fig. 5.12 an optical microscopy image of the 0.34×0.34 mm heater area. When the resistive heater is a Pt spiral, the dissipation is only ~30 mW in an evacuated nanoreactor (10^{-6} mbar), and up to ~0.1 W in the presence of gas flow at 1 bar pressure. During operation, a four-point measurement monitors the resistance of the heating wire to determine the actual temperature (which is calibrated ex situ). Custom-made instrumentation amplifiers drive and read out the heater. A LabView programme not only controls these amplifiers but also stores the readings as a function of time. The feed-back time of the heating system is in the order of ms, facilitating use of the power required to keep the heater

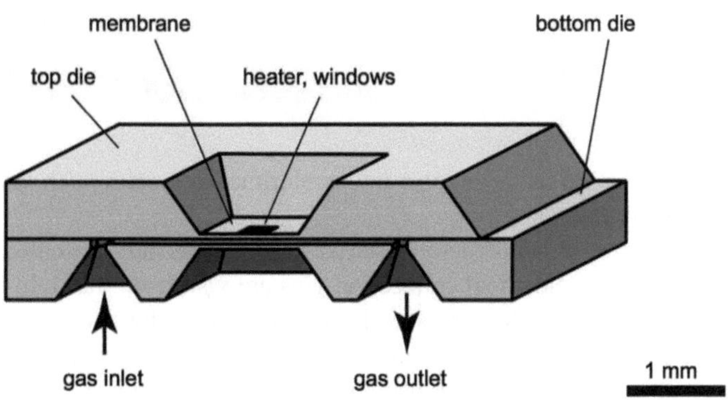

Fig. 5.10 Schematic drawing of a complete nanoreactor. From [16]

Fig. 5.11 Optical photograph of two identical nanoreactors (10 × 3.3 × 0.5 mm) seen from different angles. From [28]

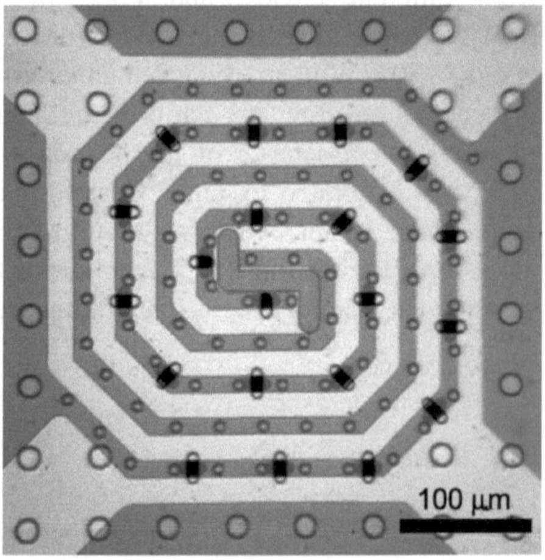

Fig. 5.12 Optical microscopy image of the heater area. From [16]

at constant temperature as an inverse microcalorimetric system such that reaction heat can be monitored.

Figure 5.13 shows a nanoreactor inserted into the specially designed and built TEM holder. The four electrical connection pins are visible on the right-hand side; the gas lines are located at the opposite side of the nanoreactor and not visible here. The complete TEM holder is shown in Fig. 5.14. The holder can be connected to a large variety of gas systems. When using only one or two unmixed gases, a simple home-built gas system can be used. However, when multiple gases need to be

Fig. 5.13 A nanoreactor inserted into the TEM gas/heating holder tip. From [16]

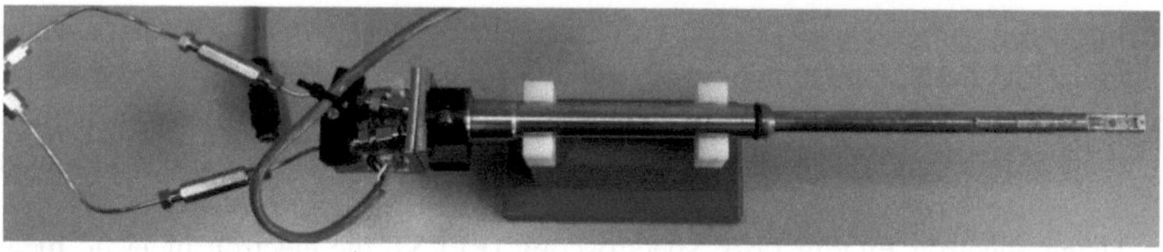

Fig. 5.14 The specially designed TEM gas flow and heating holder

mixed accurately and changes in the gas composition are also being studied, a more elaborate gas system is required. Leiden Probe Microscopy offers these gas systems [29]. Although these were originally designed to be used for *operando* scanning probe microscopy (see Chap. 1), they are very useful for TEM as well. They can handle the low flows required for the *operando* TEM experiments, provide proper handling of the exhaust gases, and exhaust gas analysis (Mass Spectrometry) can be integrated.

Overlapping silicon nitride electron transparent windows are shown in Fig. 5.15. Their thickness is 10–25 nm depending on the batch of nanoreactors. One standard silicon wafer can yield up to 50 nanoreactors, and certain specifications can be changed for each wafer if desired.

As alignment and gluing or bonding of the top and bottom wafer can be tedious procedures (possibly even leading to contamination in the TEM), a surface micro-machined process has been designed, fully integrating a nanoreactor on a single die [30]. From the outside, these nanoreactors look just like the ones shown

Fig. 5.15 TEM image of two overlapping electron-transparent windows. From [16]

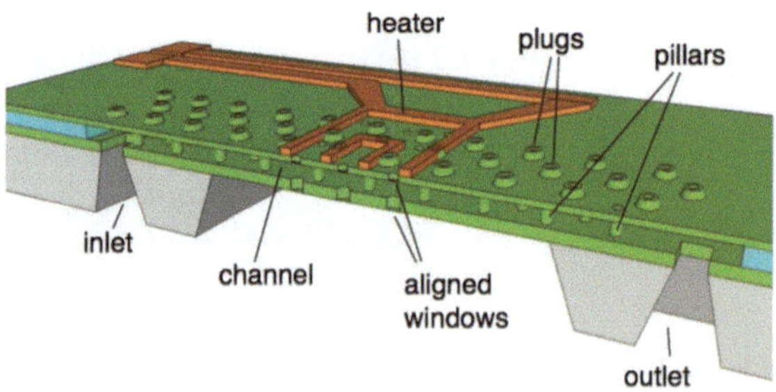

Fig. 5.16 3D-sketch of the surface micro-machined nanoreactor. From [30]

in Fig. 5.11. A sketch of this type of nanoreactor is shown in Fig. 5.16. These nanoreactors have been shown to withstand a pressure difference of at least 14 bar outside of the TEM. Because the channel height of 0.5 μm turned out to hamper loading of inorganic material, channel heights up to 5 μm have also been produced [31].

Using Electron Energy Loss Spectroscopy (EELS), the electron inelastic mean free path length through the gas confined within a nanoreactor can be determined. The distance between the two windows can be determined accurately by either measuring the Thon rings in the diffractogram of a defocused image of the two amorphous windows, or by focusing first on one window and then on the other and determining the difference in defocus value. As the inelastic mean free path of an electron within a given gas is proportional to the density of that gas, it can be used to determine the gas density. Once the gas density and the thickness of the gas layer are known, the ideal gas law can be used to either determine the local gas pressure when the temperature is known [24], or to determine the local temperature when the gas pressure is known [32]. The latter paper elegantly shows that the centre of the heater area is warmer than its edge, which can be explained by energy dissipation. This is illustrated in Fig. 5.17.

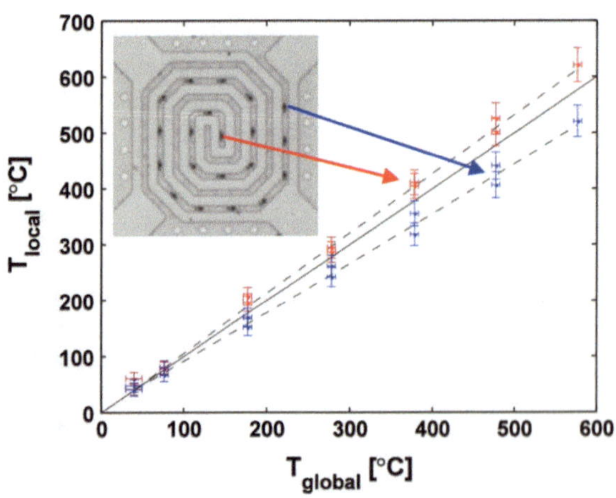

Fig. 5.17 The local temperature based on EELS measurements at the centre (*red*, upper data points) and edge (*blue*, lower data points) windows of the heated area of a nanoreactor as a function of the global temperature from the resistivity measurement. The *solid black line* has a slope of one and indicates perfect coincidence for the local and global temperature. From [32]

5.4 Towards *Operando* Catalysis

The nanoreactor system has so far been used to study hydrogen storage [33], localised corrosion in an aluminium alloy [34], catalyst preparation, and several catalytic processes. Many more catalytic studies are in progress at the moment and will be published as soon as sufficient data have been recorded and analysed.

When performing *operando* TEM, up to five different computers are used simultaneously. One computer regulates and monitors the heater temperature and power input, one computer regulates and monitors the gas system and exhaust gas analysis, one computer operates the TEM and one or two computers are used for TEM image capturing. In order to interpret dynamic data properly, either the internal clocks of all these computers need to be synchronised exactly or the scientist spends a lot of time matching data from different computers manually. To integrate the complete dataset acquired by all these computers into a single, unified view, and to help establish correlations between the TEM images and the other signals, the open-source software tool Spacetime has been developed [35].

The first relevant data using the nanoreactor system have been recorded for the reduction of copper oxide on a zinc oxide support [16, 27]. Metallic copper on zinc oxide is a well-known catalyst for methanol synthesis and the water-gas-shift reaction, but as metallic copper is not stable in ambient air it needs to be prepared in situ from an air-stable oxidic precursor. In addition to being a reaction that requires both a protective atmosphere (hydrogen gas) and heating, the reduction of zinc-oxide-supported copper oxide also provides a very suitable system for testing the resolution of the nanoreactor during TEM imaging. The Cu(111) and Cu(200) lattice planes have d-spacings of 0.21 nm and 0.18 nm, respectively. For zinc oxide these values are 0.24 nm for (101) and 0.28 nm for (100). The reduced copper on zinc oxide thus shows a set of d-spacings, as is illustrated in Fig. 5.18. The image of Fig. 5.18a was recorded at 500 °C in 1.2 bar of hydrogen gas and still clearly shows the Cu(111) lattice planes, proving the resolution of the *operando* system to be at least 0.18 nm. The corresponding Fourier transform of Fig. 5.18b confirms at least 0.18 nm resolution.

A phenomenon that is of interest in catalyst regeneration is the Kirkendall effect. It concerns the transformation of metal nanoparticles into hollow metal oxide nanoparticles. The void formation is attributed to different diffusivities of metal and oxygen atoms through the oxide layer covering the metal nanoparticle surface: Because metal atoms diffuse faster outwards through the oxide layer than oxygen atoms diffuse inwards, vacancies accumulate in the core of the nanoparticle and condense into a void in order to minimise the free energy. This mechanism has been deduced based on traditional before-and-after TEM. A few dynamic studies on the Kirkendall effect in various metals have already been published using ETEM (see for example for Ni [36]). More recently, a study on the Kirkendall effect during the oxidation of cobalt nanoparticles was published [23]. This study shows the formation of the hollow cores and quantifies the hollow core volume. The total pressure used is not explicitly stated, but can be inferred to be 1.1 bar from other

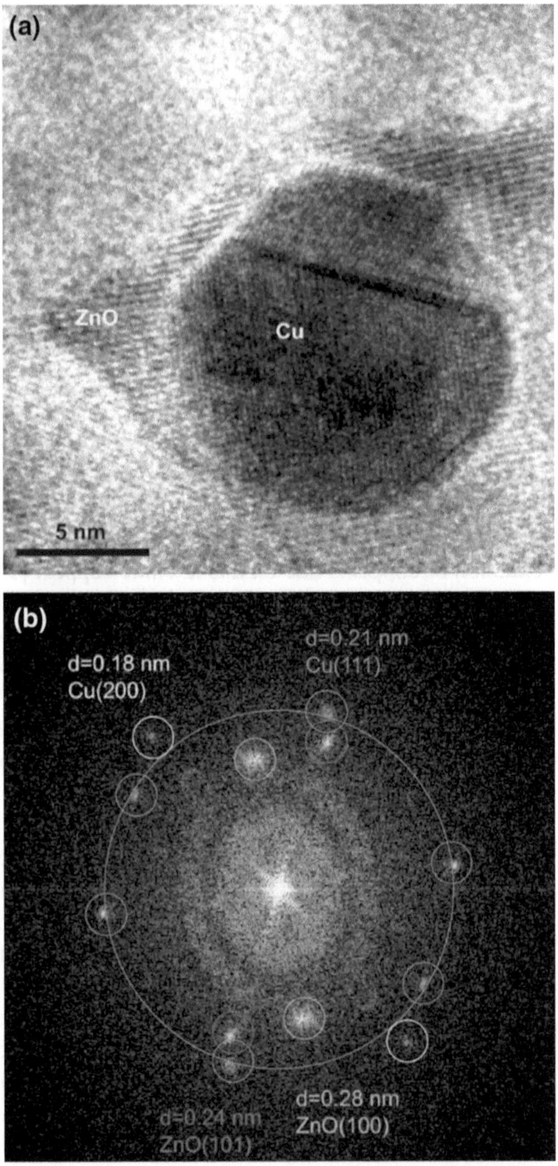

Fig. 5.18 **a** An *operando* reduced copper particle on zinc oxide recorded in 1.2 bar hydrogen at 500 °C. The lattice fringes (*parallel lines*) of the metallic copper and of the zinc oxide are clearly visible. **b** The Fourier transform of the image shown in (**a**). The *bright dots* represent sets of lattice fringes. The large concentric *circle* corresponds to a spacing of 0.21 nm. The smallest resolved lattice spacing is 0.18 nm. From [16]

experiments described in the same paper; temperatures used are up to 350 °C. Figure 5.19 shows consecutive images of three individual particles during heating at 150–250 °C. The growth of the hollow cores is visible, but high-resolution images have only been obtained after complete oxidation.

We have recently performed high-resolution TEM studies of the Kirkendall effect during the oxidation of copper nanoparticles [37] in 50 mbar of oxygen gas

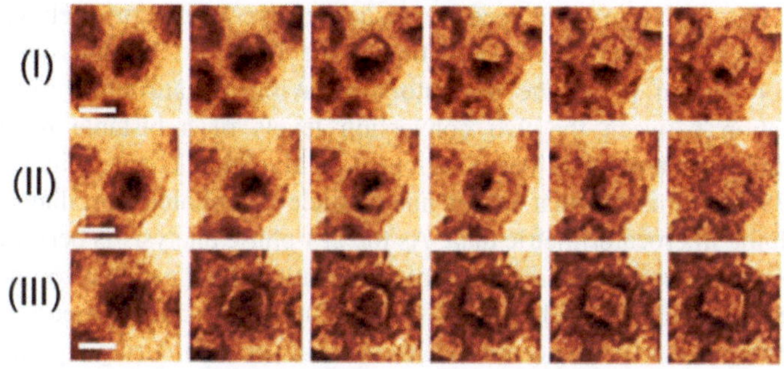

Fig. 5.19 Consecutive TEM images of three individual cobalt nanoparticles recorded in situ during oxidation in 1.1 bar of oxygen at 150–250 °C. *Scale bar* is 10 nm. From [23]

using the nanoreactor system. Even at this low pressure the copper nanoparticles are immediately covered in a thin layer of oxide at room temperature, and become completely oxidised at 150 °C. The irregular growth of the oxide shell coupled to the void formation starts after a few minutes at 150 °C, but changes in particle radius are already visible just before reaching 150 °C. This is illustrated in Fig. 5.20.

As carbon monoxide, CO, is poisonous both to mammals and certain catalysts, its oxidation to relatively harmless carbon dioxide, CO_2, is an important reaction. As a result, it has already been studied using a wide array of different techniques, from bulk catalysis to high-vacuum surface-science techniques [38]. Both in bulk catalysis as well as in model surface-science reactions, spontaneous oscillations in the reaction rate under seemingly fixed reaction conditions have been observed and

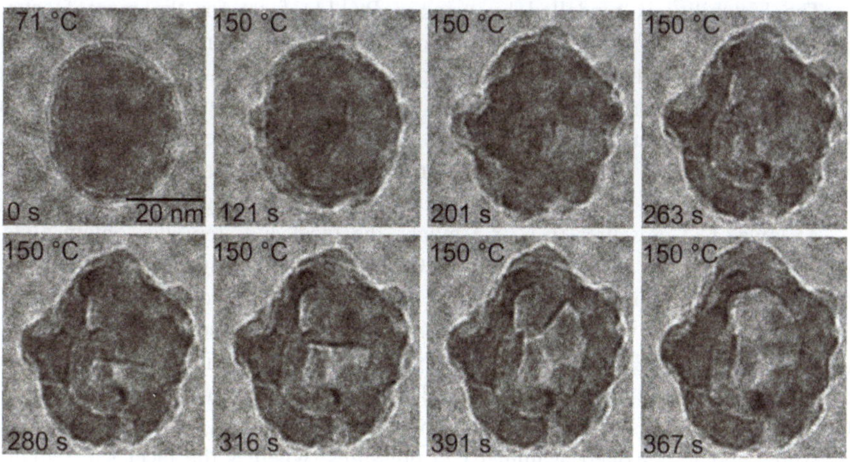

Fig. 5.20 Consecutive TEM images of a copper nanoparticle recorded in situ in a nanoreactor during oxidation in 50 mbar of oxygen

described [39, 40]. We have recently visualised the oscillatory behaviour of Pt nanoparticles catalysing the CO oxidation in the TEM at 1 bar pressure using the nanoreactor system [41]. The oscillatory behaviour is attributed to a periodic transformation between the two conversion levels of a bistable reaction. The periodic transformation has been attributed to dynamic changes in the catalyst surface [39]. It turns out that the Pt nanoparticle morphology oscillates synchronised with the microcalorimetric data. As the gas-phase analysis is performed using a mass spectrometer at some distance from the outlet of the nanoreactor, the exhaust gas-phase composition oscillates with a different phase: During a period of oscillatory behaviour the CO concentration is much lower than during a period without oscillatory behaviour. The length of the gas lines between the nanoreactor outlet and the mass spectrometer means the gas composition is levelled off by space-time broadening: During transport from the nanoreactor exit to the mass spectrometer entrance, the gases are mixed to an uncontrollable and hard-to-determine extent. This means that the composition of the gas once it reaches the mass spectrometer entrance is averaged over time with respect to the composition as it exits the nanoreactor.

Figure 5.21 shows the gas composition, the nanoreactor heater power required to maintain a stable temperature, the Pt nanoparticle shape factor, and the Pt nanoparticle itself, all during the same typical period of oscillatory behaviour. As is also observed in bulk catalysis, periods of oscillatory behaviour arise spontaneously during a period of apparently stable reaction behaviour, but can also be triggered by a small increase in temperature of 0.5 or 1 °C. The dynamic refacetting has also been studied at atomic resolution during the same set of experiments. This is shown in Fig. 5.22. The Pt(111) lattice spacings are visible in the images themselves, and also show in the Fourier transforms (insets). The orientation of the observed Pt(111) lattice fringes is consistent with the superimposed crystal lattice vectors and zone axis (Z.A.).

The bistability is modelled by using a Pt(111) facet for the more facetted morphology and a stepped Pt(211) facet for the more rounded morphology, creating a two-site microkinetic model [41]. At high CO pressures ($p_{CO} > 1$ mbar), the step is almost fully covered by CO, whereas the facet is only sparsely covered. The surface sites are practically depleted in O. The step is therefore stabilised compared to the facet, because CO adsorption is stronger at the step than at the facet, and a more rounded shape of the Pt nanoparticles is expected at high CO pressure.

At lower CO pressures ($p_{CO} < 1$ mbar), the facet and step sites obtain a significant O coverage. As the bonding of O is similar in strength at both sites, the facet is stabilised compared to the step, and a more facetted shape is expected. Thus, in agreement with [42], the morphology change of the Pt nanoparticles is attributed to the CO pressure variation under the reaction conditions. The gas-dependent shape of the Pt nanoparticles affects the CO conversion, because the CO oxidation rate is site-dependent. The step site is more active at lower CO pressures and is poisoned at higher CO pressures, whereas the facet site remains more constant in activity over the CO-pressure range used in the experiment. A dynamic and reversible refacetting of Pt nanoparticles represents a mechanism

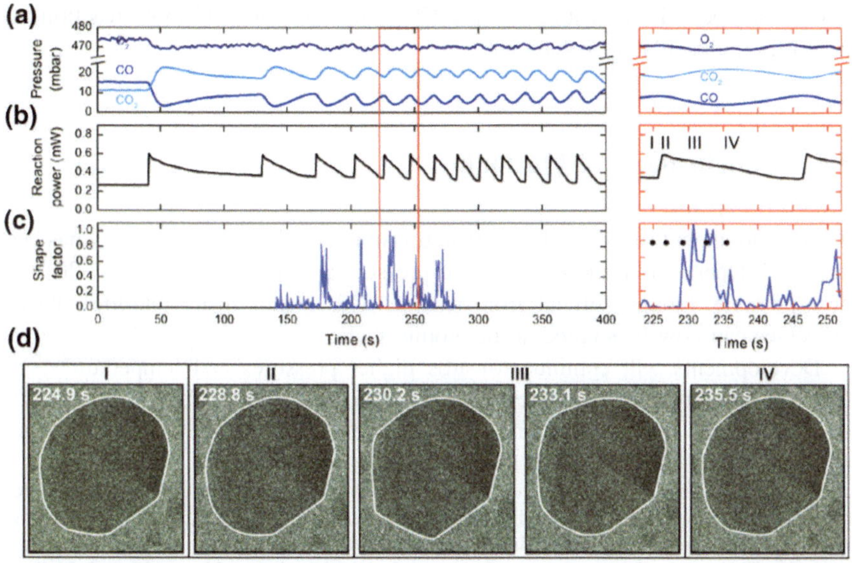

Fig. 5.21 Correlation of oscillatory CO oxidation reaction data with the projected morphology of a Pt nanoparticle. The gas entering the reaction zone is 1.0 bar of 3%:42%:55% CO:O_2:He; nanoreactor temperature is 368 °C. **a** Mass spectrometry of the CO, O_2 and CO_2 pressures. **b** Reaction power. **c** Shape factor. **d** Time-resolved TEM images of a Pt nanoparticle at the gas exit of the reaction zone. The shape factor corresponds to the relative difference in area from the best elliptical fit in image I in (**d**). Part of the reaction oscillation data for (**a–c**) is highlighted by the *red rectangle* and given with a more detailed time axis on the *right-hand side*. From [41]

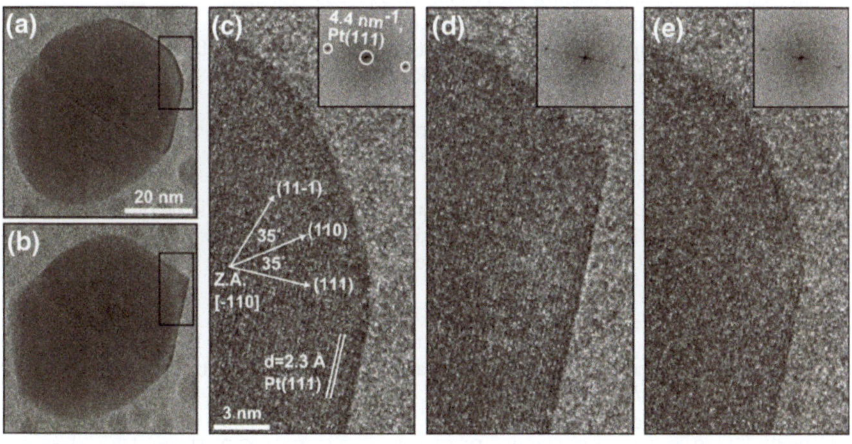

Fig. 5.22 Time-resolved, high-resolution TEM images of a Pt nanoparticle at the gas exit of the reaction zone. The gas entering the reaction zone is 1.0 bar of 4.2%:21.0%:74.8% CO:O_2:He; nanoreactor temperature is 454 °C. **a–e** The TEM images show the more spherical shape (**a**, **c**, **e**) and the more facetted shape (**b**, **d**), during the oscillatory reaction. Fast Fourier transforms are included as *insets* in (**c–e**). From [41]

that causes periodic transitions in the CO conversion in a bistable reaction and describes the oscillatory behaviour.

5.5 Conclusion and Outlook

The development towards real *operando* transmission electron microscopy has been achieved. Operation at more than 1 bar pressure and heating to at least 600 °C is now possible, though still far from routine. This means many real-life catalytic processes can now be studied at the atomic scale.

Developments will continue towards higher pressures and temperatures, and towards more routine equipment. It will also become possible to study more reactive species such as acids and corrosive gases.

Acknowledgements This chapter would not have been written without the practical and conceptual work performed by and in collaboration with Henny W. Zandbergen, J. Fredrik Creemer, Joost W.M. Frenken, Gert-Jan van Baarle, G. Marien Bremmer, Peter A. Crozier, Pleun Dona, Christian Elkjær, Stig Helveg, Stephan Janbroers, Luigi Mele, Bruno Morana, Bart J. Nelissen, (Pita) I. Puspitasari, Sander B. Roobol, Richard van Rijn, Søren B. Vendelbo, and Sven Ullmann. The work on the nanoreactor system was performed in the framework of NIMIC (Nano IMaging under Industrial Conditions), a SmartMix project of the Dutch Ministry of Economic Affairs [26].

References

1. M. Knoll, E. Ruska, Zeitschrift für Physik **78**, 318–339 (1932)
2. F. Wu, N. Yao, Nano Energy **13**, 735–756 (2015)
3. F.E.I. company. http://www.fei.com/products/tem/titan-etem/?ind=MS
4. P.J. Ferreira, K. Mitshuishi, E.A. Stach, MRS Bull. **33**, 83–90 (2008)
5. P. Gai, Top. Catal. **8**, 97–113 (1999)
6. H. Hashimoto, T. Naiki, T. Eto, K. Fujiwara, Jpn. J. Appl. Phys. **7**, 946–952 (1968)
7. R.T.K. Baker, Catal. Rev.: Sci. Eng. **19**, 161–209 (1979)
8. R.T.K. Baker, Carbon **27**, 315–323 (1989)
9. Haldor Topsøe company. http://topsoe.ru/research/Resources/Microscopy_movies.aspx?movieid={59D9E277-29E3-41A0-B49F-50519295C89E}
10. S. Janbroers, P.A. Crozier, H.W. Zandbergen, P.J. Kooyman, Appl. Catal. B **102**, 521–527 (2011)
11. F. Tao, P.A. Crozier, Chem. Rev. **116**, 3487–3539 (2016)
12. T.W. Hansen, J.B. Wagner, Controlled Atmosphere Transmission Electron Microscopy—Principles and Practice. Springer International Publishing Switzerland (2016). ISBN 978-3-319-22987-4/22988-1
13. A. Kruize, Private communication
14. S. Giorgio, S. Sao Joao, S. Nitsche, D. Chaudanson, G. Sitja, C.R. Henry, Ultramicroscopy **106**, 503–507 (2006)
15. P.L. Gai, E.D. Boyes, S. Helveg, P.L. Hansen, S. Giorgio, C.R. Henry, MRS Bull. **32**, 1044–1050 (2007)
16. J.F. Creemer, S. Helveg, G.H. Hoveling, S. Ullmann, A.M. Molenbroek, P.M. Sarro, H.W. Zandbergen, Ultramicroscopy **108**, 993–998 (2008)

17. Protochips. http://www.protochips.com/products/atmosphere-gas-environmental-cell.html
18. L.F. Allard, W.C. Bigelow, S.H. Overbury, D.P. Nackashi, J. Damiano, Microsc. Microanal. **16**, 296–297 (2010)
19. L.F. Allard, S.H. Overbury, W.C. Bigelow, M.B. Katz, D.P. Nackashi, J. Damiano, Microsc. Microanal. **18**, 656–666 (2012)
20. L.F. Allard, W.C. Bigelow, S. Zhang, X. Pan, Z. Wu, S.H. Overbury, W.B. Carpenter, F.S. Walden, R.L. Thomas, D.S. Gardiner, B.W. Jacobs, D.P. Nackashi, J. Damiano, Microsc. Microanal. **20**, 1572–1573 (2014)
21. Hummingbird Scientific. http://hummingbirdscientific.com/products/gas-flow/
22. D.H. Alsem, R.R. Unocic, G.M. Veith, K.L. More, N.J. Salmon, Microsc. Microanal. **18**, 1158–1159 (2012)
23. H.L. Xin, K. Niu, D.H. Alsem, H. Zheng, Microsc. Microanal. **19**, 1558–1568 (2013)
24. R. Colby, B. Kabius, D.H. Alsem, Microsc. Microanal. **19**, 474–475 (2013)
25. http://DENSsolutions.com/products/gas
26. NIMIC. http://nimic.physics.leidenuniv.nl
27. J.F. Creemer, S. Helveg, G.H. Hoveling, S. Ullmann, P.J. Kooyman, A.M. Molenbroek, H.W. Zandbergen, P.M. Sarro, in IEEE International Conference on MEMS Technical Digest, vol. 22 (2009), pp. 76–79
28. J.F. Creemer, S. Helveg, P.J. Kooyman, A.M. Molenbroek, H.W. Zandbergen, P.M. Sarro, J MEMS **19**, 254–264 (2010)
29. Leiden Probe Microscopy. http://www.leidenprobemicroscopy.com
30. J.F. Creemer, F. Santagata, B. Morana, L. Mele, T. Alan, E. Iervolino, G. Pandraud, P.M. Sarro, M.E.M.S. Proceedings, Cancun. MEXICO **2011**, 23–27 (2011)
31. B. Morana et al., In preparation
32. S.B. Vendelbo, P.J. Kooyman, J.F. Creemer, B. Morana, L. Mele, P. Dona, B.J. Nelissen, S. Helveg, Ultramicroscopy **133**, 72–79 (2013)
33. T. Yokosawa, T. Alan, G. Pandraud, B. Dam, H. Zandbergen, Ultramicroscopy **112**, 47–52 (2012)
34. S. Malladi, C. Shen, Q. Xu, T. de Kruijff, E. Yücelen, F. Tichelaar, H.W. Zandbergen, Chem. Commun. **49**, 10859–10861 (2013)
35. Roobol, S.B., in preparation, http://spacetime.uithetblauw.nl
36. S. Chenna, P.A. Crozier, Micron **43**, 1188–1194 (2012)
37. S.B. Vendelbo et al., In preparation
38. H.J. Freund, G. Meijer, M. Scheffler, R. Schlögl, M. Wolf, Angew. Chem. Int. Ed. **50**, 10064–10094 (2011)
39. R. Imbihl, G. Ertl, Chem. Rev. **95**, 697–733 (1995)
40. M.A. Liauw, P.J. Plath, N.I. Jaeger, J. Chem. Phys. **104**, 6375–6386 (1996)
41. S.B. Vendelbo, C.F. Elkjaer, H. Falsig, I. Puspitasari, P. Dona, L. Mele, B. Morana, B. J. Nelissen, R. van Rijn, J.F. Creemer, P.J. Kooyman, S. Helveg, Nat. Mater. **13**, 884–890 (2014)
42. H. Yoshida, K. Matsuuara, Y. Kuwauchi, H. Kohno, S. Shimada, M. Haruta, S. Takeda, Appl. Phys. Express **4**, 065001 (2011)

Chapter 6
Planar Laser Induced Fluorescence Applied to Catalysis

Johan Zetterberg, Sara Blomberg, Jianfeng Zhou, Johan Gustafson and Edvin Lundgren

Abstract In this chapter we describe Planar Laser Induced Fluorescence (PLIF) to investigate the reactants or products in the vicinity of a catalyst at semi-realistic conditions. PLIF provides a 2D view of the gas-phase distribution of a pre-chosen gas. Here we present PLIF results from CO_2 and CO from the oxidation of CO into CO_2 by Pd single crystals and by various powder catalysts as well as from NH_3 from the oxidation of NH_3 above a Ag/Al_2O_3 powder catalyst. We describe our experimental set-up in detail, and the laser instrumentation needed to enable detectable gas fluorescence from CO_2, CO, and NH_3, respectively. Further, intensity corrections of the PLIF signal due to scattering and temperature effects are described. In the case of the CO oxidation, the results directly show the creation of a CO_2 boundary layer and thus a drastic change of the gas-phase composition close to the catalyst surface, illustrating the effect of gas diffusion and reaction speed, which in turn should affect the surface structure of the active catalyst. The 2D character of the PLIF images is used to investigate differences in catalyst activity by studying adjacent catalysts in the reaction cell during the reaction, and a solution to avoid spill-over effects between catalysts in the same reactor is presented. The results from PLIF images of CO of the same reaction show the corresponding depletion of the PLIF intensity above the catalyst, in accordance with observations from other techniques confirming the drastic difference between the gas composition close to the catalyst and at the inlet or outlet of the reactor. Finally we present NH_3 PLIF results from above a Ag/Al_2O_3 powder catalyst while the NH_3 is being oxidized in an oxidizing environment with the assistance of H_2.

J. Zetterberg · J. Zhou
Division of Combustion Physics, Lund University, 221 00 Lund, Sweden

S. Blomberg · J. Gustafson · E. Lundgren (✉)
Division of Synchrotron Radiation Research, Lund University, 221 00 Lund, Sweden
e-mail: edvin.lundgren@sljus.lu.se

J. Frenken and I. Groot (eds.), Operando *Research in Heterogeneous Catalysis*,
Springer Series in Chemical Physics 114, DOI 10.1007/978-3-319-44439-0_6

6.1 Introduction

Laser Induced Fluorescence (LIF) is a common, non-intrusive laser diagnostic technique for gas-phase studies in a number of research fields [1, 2] but is less known to the catalytic community. By the use of LIF, temperature, velocity and concentration of the probed gas can be determined [3]. The gas is probed with a laser tuned to a wavelength that matches an energy-level transition of the molecule of interest. If the wavelength is correctly chosen, no other molecular species will have an overlapping transition at that specific energy and the absorption cross section for the transition is high. When the molecule relaxes, fluorescent light is emitted which can be detected. This results in a strong LIF signal coming from one specific species, making concentration measurements on sub-ppm levels possible and LIF has therefore been used extensively in the investigation of the gas composition in flames and in other combustion-related environments.

In contrast, the use of LIF in catalysis-related research is comparably scarce. Previously, LIF has been used to study products or reactants such as OH [4–7], formaldehyde [8, 9], or naphthoquinone [10], in catalytic reactions. These molecules are easily accessible in the ultraviolet/visible region, making excitation and detection straightforward. However, a number of the more interesting molecules for catalysis such as CO, CO_2, and small hydrocarbons fluoresce in the mid-infrared region, making both excitations and detection more complicated. For instance, to detect CO_2 at elevated temperatures and realistic gas conditions, it is necessary to excite the molecule with a wavelength of 2.7 μm, demanding a high-energy laser with a respectable power and narrow line width (energy distribution). Furthermore, a fast, gateable infrared camera with the ability to detect the fluorescent light at 4.3 μm is necessary, since a fast subtraction scheme is needed to remove the otherwise completely dominating thermal background. Thus both laser source and detector camera are non-standard, making the experimental combination and the presented results in the present chapter rare or even unique. To obtain a 2D image of the gas-phase distribution above a catalyst, a laser sheet can be created, and the technique is then called Planar Laser Induced fluorescence (PLIF).

Motivated mainly by the difficulties in determining the active surface phase for Pd and Rh model catalysts during CO oxidation under semi-realistic conditions [11–24] and theoretical calculations [25–29], we have during recent years developed an experimental set-up suitable for probing the gas phase above the surface of an active catalyst using PLIF [30]. A simplified illustration of the PLIF experiment applied to catalysis is shown in Fig. 6.1 for the case of CO oxidation and detection tuned for CO_2, and the goal is to measure the amount of a reactant or product in the vicinity of the catalyst as we are changing the nature of the catalyst or the environments such as the gas pressure or sample temperature.

In the present chapter we describe our experimental set-up including the reactor and the gas flow system. We also explain which laser sources are used for a particular excitation of a molecule to induce a favorable fluorescence from the molecule of interest to enable PLIF observation of a catalytic reaction in which the molecule participates. Furthermore, the necessity of subtracting a thermal background for each

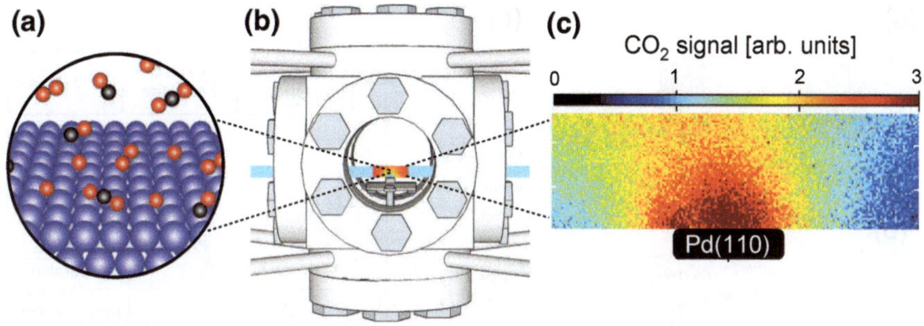

Fig. 6.1 A schematic of the PLIF measurements of the gas phase above a catalyst as presented in this chapter. **a** The adsorption and dissociation of O_2 and the adsorption of CO and production of CO_2. **b** The laser sheet (*blue*) that excites a vibrational-rotational level in the CO_2 molecules above the catalyst. **c** The resulting PLIF image of the CO_2 distribution above an active catalyst at elevated temperature and semi-realistic pressure

single PLIF measurement is described. In the first part we will focus on detecting the CO_2 production during CO oxidation above a single catalyst as well as above two or more catalysts probed simultaneously to demonstrate the 2D capabilities of the approach, and to possibly enable catalyst screening. These measurements also show, in a very direct way, the change of the gas-phase composition close to the catalyst surface as the catalyst becomes active. In the second part we describe how it is possible to probe the concentration of CO close to the catalyst surface during the CO oxidation and the complementary picture such measurements provide with respect to the CO_2 measurements. Finally, we will describe PLIF measurements tuned to fluorescence of NH_3 during its oxidation assisted by H_2 above a Ag/Al_2O_3 powder catalyst.

6.2 Experimental

6.2.1 Laser Set-Up and Laser Sheet

For each gas species probed in the present investigations, a unique combination of lasers and detectors was used. However, the general experimental set-up was the same in all cases, and a schematic overview is shown in Fig. 6.2a [30].

The infrared laser beam was in all cases formed into a laser sheet (6–15 mm high) by a cylindrical and a spherical lens, f = –40 mm and f = 150 mm, respectively (Fig. 6.2b), and was sent through the reactor just above the catalyst.

6.2.2 Reactor, Gas System, and Samples

In Fig. 6.2c an overall schematic of the set-up including the lasers, the mixing unit, the reactor, the IR-camera as well as the sample environmental control, for the case

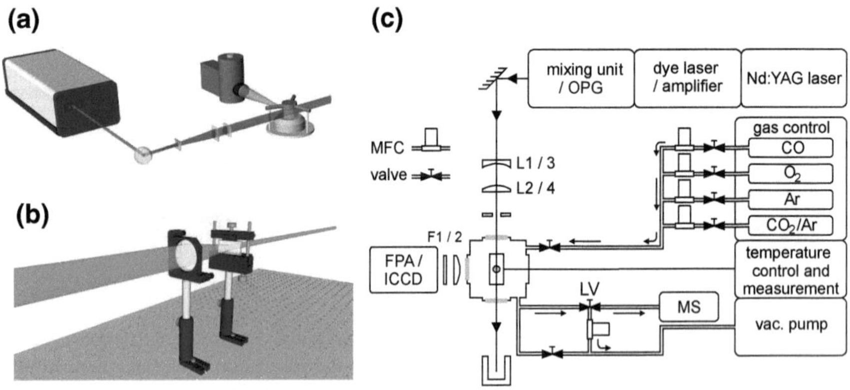

Fig. 6.2 **a** A schematic overview of the experimental set-up. **b** Illustration of transforming the point-like laser beam into a 2D sheet. **c** A more detailed description of the experimental set-up including the gas system and pumping units in the case of CO oxidation and CO_2 measurements

of CO_2 measurements, is shown. The reactor is a standard stainless-steel cube, fitted with CaF_2 windows allowing the laser to penetrate into the reactor and the fluorescence signal to reach to the detector. The samples are heated by a boron nitride heater and the surface temperature of the samples is measured by a standard type-C thermocouple and by an IR camera (FLIRP620). In the case of a single crystal, the samples were sputtered in a separate chamber before being introduced into the reactor and thus exposed to air and contaminations, while no pretreatment was performed for the powder-based samples. In all cases, the model catalysts were ramped up and down in temperature in an O_2, CO, and Ar-containing atmosphere to reduce the amount of contaminants. The feed gases were Ar, CO, O_2 and premixed 10% CO_2 in Ar for the CO_2 and CO measurements, and NH_3 and H_2 for the NH_3 measurements. The gases were introduced into the reactor cell via individual Bronkhorst mass-flow controllers (Bronkhorst EL-FLOW, 50 mln/min) that can vary the gas flow from 1 to 50 ml/min. To keep the pressure constant, a pressure controller (Bronkhorst EL-PRESS) was attached to the gas outlet. The gas composition was studied by a quadruple mass spectrometer (Pfeiffer PrismaPlus QMG220), connected to the reactor outlet via a leak valve.

6.2.3 LIF-General Considerations

The detected PLIF signal is dependent on a number of physical parameters which, in the linear excitation regime, can be described by

$$S_{LIF} = \eta_c E g f(T) \sigma_0 \chi_{\mathrm{abs}} \frac{P}{k_B T} \phi. \qquad (6.1)$$

η_c is the experimental collection efficiency, E is the laser energy, g is a function that describes the spectral overlap between the laser and the absorption spectral lineshape, $f(T)$ is the Boltzmann fraction, σ_0 is the absorption cross section of the probed gas, $\chi_{\rm abs}$ is the mole fraction of the gas that together with $P/k_{\rm B}T$ gives the gas number density, and ϕ is the fluorescence quantum yield. A more detailed description of the quantum yield related to IRLIF is given by Kirby and Hanson [31, 32]. The expression shows that the detected LIF signal is linearly dependent on the gas density and the fraction of molecules in the state from which the laser excites the molecule, $f(T)$.

As seen from the expression above, the signal is proportional to the number density of the probed gas. The number density N can be described by the mole fraction and the ideal gas law according to

$$N = \chi_{abs}\frac{P}{k_B T}. \tag{6.2}$$

This shows that the signal will decrease as a function of temperature. In order to analyze the influence of the temperature in the present measurements we measured the LIF signal at a constant pressure of 136 mbar of CO_2 while the temperature of the sample was increased from 120 to 290 °C. Figure 6.3 shows the calculated density of molecules/m^2 (dashed line) from the ideal gas law where the gas is assumed to have the same temperature as the sample together with the achieved LIF signal (solid line) 1.5 mm above the sample surface. The figure shows that the major signal decrease can be explained by the decrease of the gas number density, due to the increase of the temperature. However, the dependence on the population is not negligible and the LIF signal should be corrected for both, to represent the actual concentration distribution in the measurement.

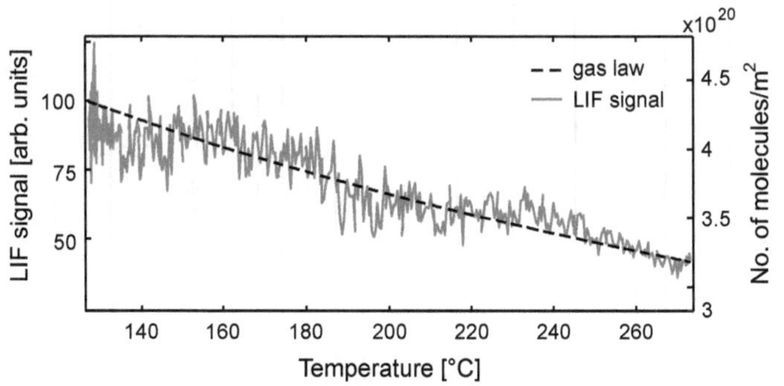

Fig. 6.3 The detected CO_2 LIF signal 1.5 mm above the sample (*full line*) as a function of temperature. The *dashed line* shows the calculated density of molecules as a function of temperature, the P(12) line was used for laser excitation

6.3 CO_2 Imaging

In Fig. 6.4a an excitation scan of CO_2 at 27 °C is shown. The present measurements were performed by exciting P12 in the $(00^00) \longrightarrow (10^001)$ transition of CO_2 as indicated in Fig. 6.4a. The energy diagram for the CO_2 molecule and the resulting fluorescence at 4.3 μm from the P12 level is shown in Fig. 6.4b.

In order to reach the P12 in the $(00^00) \longrightarrow (10^001)$ transition of the CO_2 molecule at 2.7 μm, the second harmonic at 532 nm from a Spectra Physics, PRO 290-10 Nd:YAG laser with a repetition rate of 10 Hz and an 8 ns pulse length was used to pump a Sirah PRSC-D-18 tunable dye laser using LDS 765 as dye. The residual of the fundamental beam (at 1064 nm) after the frequency doubling was difference-frequency mixed with the output of the dye laser (at 763 nm) in a $LiNbO_3$ crystal, yielding a tunable infrared laser beam at 2.7 μm, with a pulse energy of 5–8 mJ and a 5 ns pulse length. The bandwidth of the infrared laser beam was estimated in an earlier work to be 0.025 cm^{-1} [33].

Because of the fluorescence being in the infra-red region, the thermal background will interfere with the LIF signal. In Fig. 6.5 the schematics to remove the thermal background is shown as applied in the CO_2 measurements. The camera was triggered at 20 Hz, via a digital delay generator (Stanford Research Systems DG535), at every second exposure synchronized with the laser operated at 10 Hz, to subtract the background. The long radiative lifetime (100 μs) of the IR-transitions [34–36] was utilized to discriminate laser background and time jitter, by using a 15 μs time

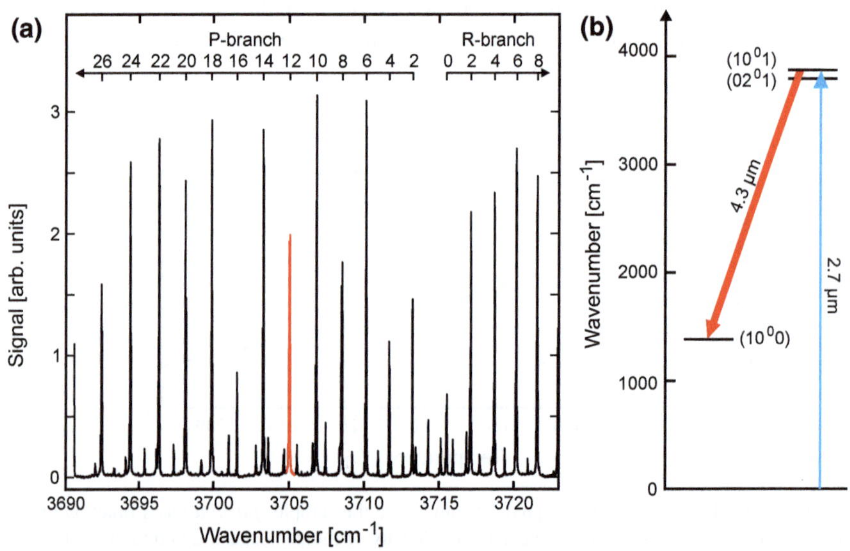

Fig. 6.4 **a** Measured rotational-vibrational excitation spectrum of CO_2 at 27 °C. The P12 is indicated in the figure and was used for excitation with a laser wavelength of 2.7 μm. **b** Energy diagram of the CO_2 molecule and the excitation by the laser sheet and the resulting fluorescence which was detected at 4.3 μm

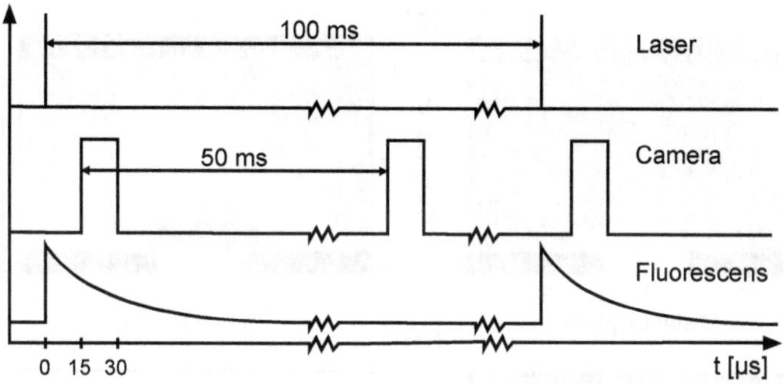

Fig. 6.5 Illustration of the time triggering scheme between the laser and the camera to remove the thermal background

delay relative to the Q-switch of the laser. In Fig. 6.5 it can be seen that the laser sets the start time and the fluorescence appears almost instantly. The camera is then triggered 15 μs after the opening of the laser Q-switch. The exposure time for the camera is optimized to discriminate background and favor fluorescence signal and set to 15 μs. The camera is then triggered once more during the cycle 50 ms after the first exposure, camera gate unchanged. This is to collect the thermal background without any influence of the laser. 100 ms after the first laser shot the next cycle starts, again with the opening of the laser Q-switch.

The fluorescence signal was collected by a CaF_2 lens at 90° to the incident laser sheet through a narrowband, liquid-nitrogen-cooled, interference filter (centered at 4.26 μm) and detected by a 256 × 256 InSb IR camera (Santa Barbara Focal Plane, SBF LP134). Using this set-up, our detection limit of CO_2 is at present 100 ppm or 0.1 mbar at a total pressure of 100 mbar.

6.3.1 One Sample: CO_2 Imaging Above a Pd(100) Single Crystal

The Pd(100) single-crystal surface is one of the most commonly used substrates as a model catalyst to study CO oxidation under semi-realistic conditions. The surface structure during the reaction has been studied with Scanning Tunneling Microscopy (STM) [11], X-ray Photoelectron Spectroscopy (XPS) [21], (High Energy) Surface X-Ray Diffraction ((HE)SXRD) [12, 24], Infrared Reflection Absorption Spectroscopy (IRAS) [15], and extensively by Density Functional Theory (DFT) calculations [25–29]. The experimental surface studies are done with simultaneous mass spectrometry data collection to relate the state of the surface with the activity of the sample, and all of the experiments are performed using reactors similar to the one described above and used in the present experiments. Despite using mass spectrometry, it is in general difficult to visualize the gas-phase distribution in such a reactor, in particular when the model catalyst becomes active.

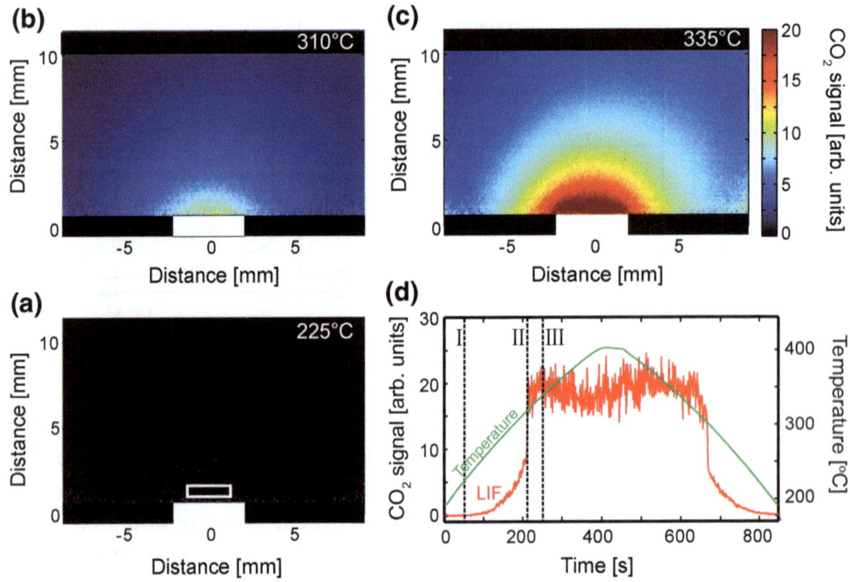

Fig. 6.6 A Pd(100) single crystal (indicated by the filled white rectangle) in a flow of 4 ml/s CO and 4 ml/s O_2 at a temperature of **a** 225 °C, **b** 310 °C, and **c** 335 °C. **d** The temperature (*green*) during the experiment and the LIF signal (*red*) extracted from the area indicated by a small non-filled rectangle

In Fig. 6.6 an experiment using a flow of 4 ml/min CO, 4 ml/min O_2 and 92 ml/min Ar through the reactor above a Pd(100) single-crystal model catalyst is shown. During the experiment, the temperature is first increased and later decreased (Fig. 6.6d) while the CO_2 LIF signal (Fig. 6.6d) from a point 1 mm above the sample is monitored. In Fig. 6.6a–c, 2D PLIF images of the CO_2 fluorescence as seen from the side of the crystal are shown at different temperatures. At a temperature of 225 °C, no signal can be detected as seen in Fig. 6.6a. However, as the temperature is increasing, a slow increase in the activity can be observed in the LIF signal. At a critical temperature, there is a sudden CO_2 increase, and 2D PLIF images at the transition are shown in Fig. 6.6b, c. In Fig. 6.6c a significant amount of CO_2 is observed, directly showing the formation of a CO_2 cloud in the form of a so-called boundary layer. As the temperature is increased further, the CO_2 production is constant at a high value. From these images alone it is evident that the gas-phase composition in the vicinity of the Pd(100) surface is very different from the gas composition introduced by the mass flow controllers, and as a result, the surface composition of the model catalyst is not reflected by the introduced gas composition, but rather by the local gas composition close to the catalyst surface. Furthermore, the CO_2 production becomes constant despite increasing the temperature. The reason for this is that, under these conditions, the oxidation of the CO at the surface is faster than the CO flow to the surface, leading to the build-up of the CO_2 boundary layer. It is said that the catalyst

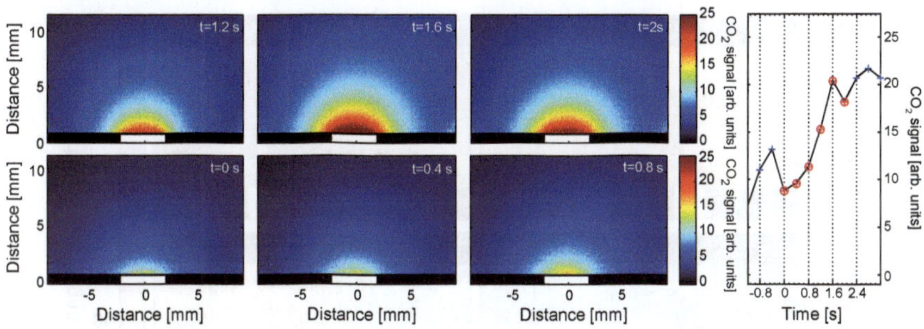

Fig. 6.7 CO_2-tuned PLIF images recorded with a 0.4 s resolution above a Pd(100) single crystal as the reaction ignites. Any region of interest in the images can be plotted as a function of time. The *right panel* shows the LIF intensity just above the sample

is in a Mass Transfer Limited (MTL) regime, which can also be observed by mass spectrometry and other methods [13, 15, 21]. However, the visual 2D information that the PLIF imaging provides generates a new and complementary experimental view and a direct understanding of this phenomenon.

The observation of a sudden change in the CO_2 production clearly suggests a change in the reaction pathway at the surface, suggesting a change of the adsorbed molecules or in the surface structure. The most likely scenario is the desorption of CO and/or a resulting oxidic (chemisorbed/oxide) layer on the surface.

The relatively high time resolution (see Fig. 6.7) of the PLIF imaging enables the extraction of the CO_2 concentration at an arbitrary position in the 2D images and to simultaneously follow the CO_2 change when conditions such as the temperature or the gas composition are changing, as long as the corrections of the LIF intensity as described above are taken into account. Each image in Fig. 6.7 is an average of four consecutive PLIF images with 0.1 s resolution. For instance, with a proper treatment of the background levels in the images, it is possible to extract Arrhenius plots to obtain activation energies for a certain catalyst under investigation.

6.3.2 *Two Samples*

An attractive property of the PLIF measurements as performed in the present experiments is the 2D character of the imaging. As a result, a spatial resolution of the gas phase of around 0.4 mm can easily be achieved, and therefore it is possible to study inhomogeneous samples or to directly compare the activity of two different samples.

In Fig. 6.8 two different powder catalysts, indicated by the white filled squares, are studied in the same way as the single-crystal Pd sample shown above. The right-hand sample contains 2% (by weight) of Pd in CeO_2 while the left hand sample contains 2% of Pt and 1% of Pd also in CeO_2 as indicated in the bottom panel of Fig. 6.8a. The samples are separated by approximately 8 mm. By taking images with a high time resolution and increasing the temperature continuously of both samples

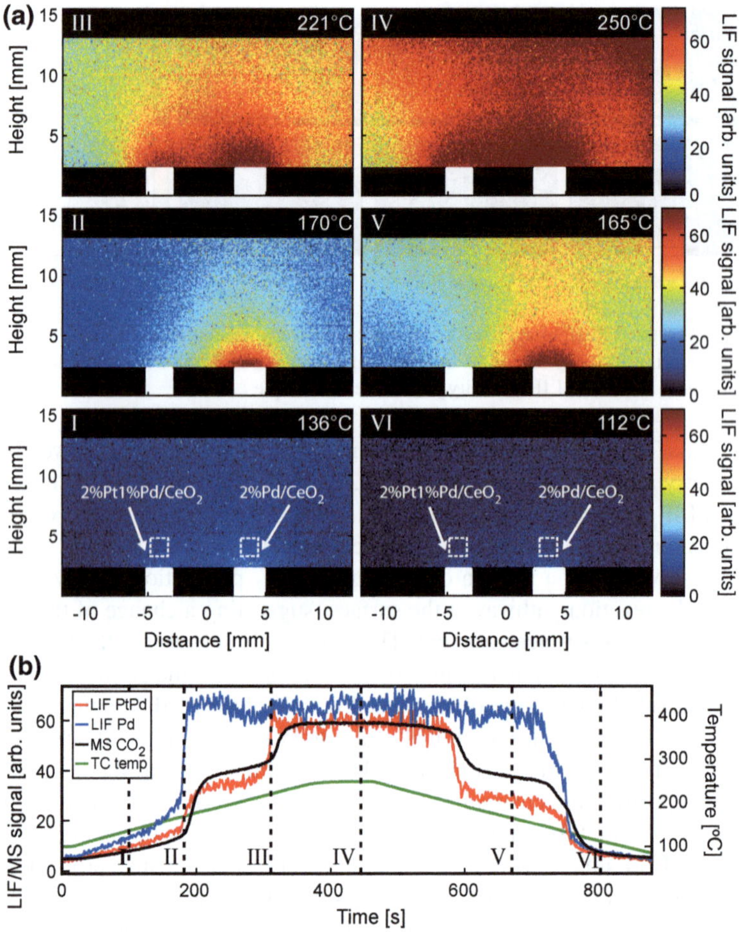

Fig. 6.8 A CO oxidation experiment imaging the CO_2 fluorescence above two powder catalysts (2%Pt1%Pd/CeO_2 and 2%Pd/CeO_2) in the reactor. **a** PLIF images collected at increasing temperatures (*I–III*) and decreasing temperatures (*IV–VI*). It is immediately clear that the 2%Pd/CeO_2 starts to produce CO_2 at a lower temperature than the 2%Pt1%Pd/CeO_2 sample. **b** The LIF signal from the indicated non-filled white squares in Fig. 6.8a *I* and *VI*, as well as the temperature. The mass spectrometer signal has been added for comparison

simultaneously in the same gas flow, we can evaluate the difference in activity of the samples. From the 2D images in Fig. 6.8a, showing snapshots during the experiment, it is directly clear that the 2%Pd/CeO_2 powder sample reaches the MTL, and is thus highly active, at a temperature which is approximately 55 °C lower than for the 2%Pt%Pd/CeO_2 sample. The CO_2 mass spectrometry signal overlaid on the PLIF signals in Fig. 6.8b has a similar, but smeared-out, profile as the PLIF signal from the 2%Pt%Pd/CeO_2 sample. The reasons for the lower activation temperature for the 2%Pd/CeO_2 powder sample could be many, and would require structural characterization of the catalysts, which is beyond the scope of the present chapter, in which we instead focus on the capabilities of PLIF. Nevertheless, the measurements in Fig. 6.8a show an interesting potential for direct comparison between the catalytic properties of different catalysts for screening purposes.

However, it is also clear from the 2D images, and in particular the point measurements directly above the catalysts as shown in Fig. 6.8b, that there is a spill-over effect from the 2%Pd/CeO_2 powder sample as it becomes active to the 2% Pt%Pd/CeO_2 sample. In experiments performed for the same samples but only one sample at a time, while no difference for the 2%Pd/CeO_2 sample was observed, the 2%Pt%Pd/CeO_2 became highly active and deactivated at temperatures different from when being in the presence of the 2%Pd/CeO_2 sample. Thus, we conclude that the gas diffusion generated by the sample with the lowest activation temperature affects the sample which becomes active at higher temperature.

6.3.3 Three Tubes

For practical purposes and if high accuracy is needed in determining differences in catalytic activity, it is necessary to separate the catalysts to avoid any spill-over effects. Such an experiment was performed by probing the exhaust from three separated glass-tubes inside the reactor chamber. In the glass-tubes, different catalysts were placed and the gas was directed to flow through the three tubes. The tubes ended inside the middle of the vacuum chamber where the laser sheet was placed as close as possible (<1 mm) to the exits of the tubes. The gas flow was kept the same through all three tubes, and the temperature was increased equivalently by a heating tape around the tubes, and measured by a thermocouple situated between the tubes. This solution enables an independent measurement of the CO_2 production from each catalyst under the same conditions.

The results are shown in Fig. 6.9 for powder catalysts consisting of 2%Pd/CeO_2, 2%Pt/CeO_2 and 2%Pt/Al_2O_3 dispersed on monoliths with the same diameter as the glass tubes. At low temperature, none of the catalysts produces CO_2 as can be seen in Fig. 6.9a, panel I. However, as the temperature is increased, the catalysts become active in CO_2 production, starting with the Pd/CeO_2 at 180 °C, the Pt/CeO_2 at 280 °C, and the Pt/Al_2O_3 catalyst at 350 °C, in Fig. 6.8a panels II, III, and IV, respectively. The mass spectrometry also detects changes in the CO_2 production but is unable to detect which of the samples becomes active at a certain temperature. In this case it is possible to extract the activation energies from the LIF signal by producing an Arrhenius plot from the development of the CO_2 production at the outlet of each tube.

The above experiment using three exhaust tubes separating product gas from each sample is an example on how PLIF may be used to discriminate between different catalysts. The example enables a direct view of the amount of the products at a given temperature. It is, however, clear that the experimental set-up can be modified and simplified to a large extent, and may have potential for practical use.

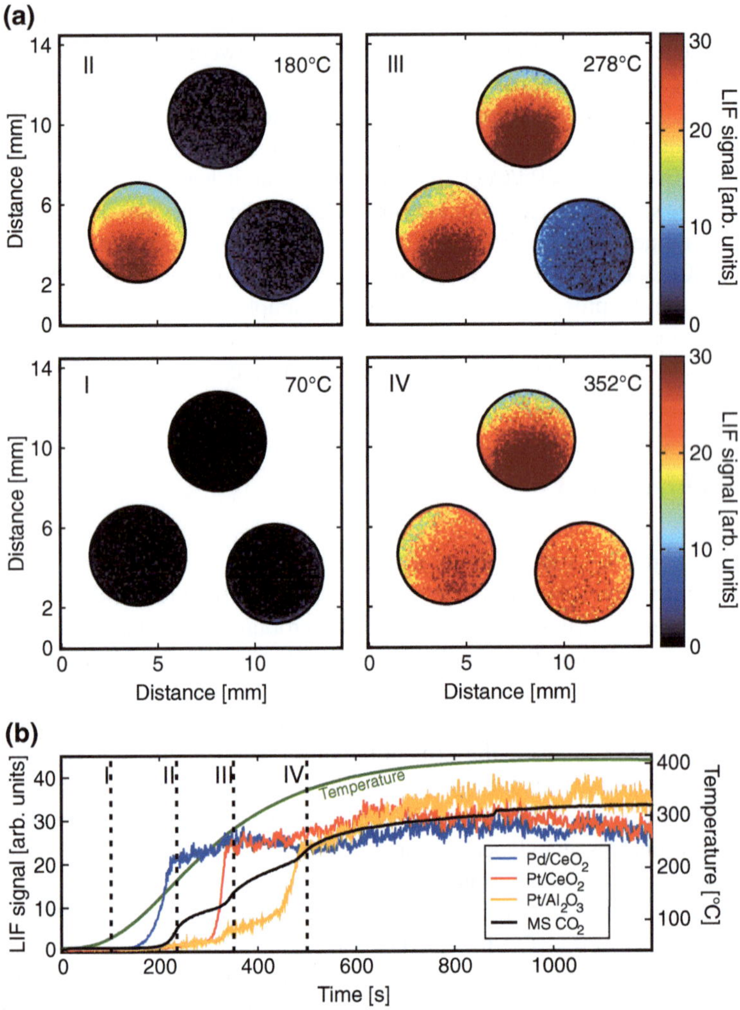

Fig. 6.9 An experiment using three exhaust tubes containing active CO oxidation powder catalysts (Pd/CeO_2, Pt/CeO_2 and Pt/Al_2O_3) exiting in the vacuum PLIF reactor imaging the CO_2 fluorescence just above the tubes simultaneously. **a** PLIF images as the temperature of the tubes is increasing from 70 to 352 °C. **b** Extracted LIF intensity at the outlet of each tube as a function of temperature. The CO_2 mass spectrometry signal is added as a reference

6.4 CO Imaging

Electronic CO resonances are located in the VUV regime and LIF is obtained via two-photon excitation in the $B^1\sum^+ \longleftarrow X^1\sum(0,0)$ Hopfield-Birge band using 230 nm (see Fig. 6.10a [2]), followed by population of the A-state via collisions or radiative transitions with fluorescence emission bands in the wavelength range 450–

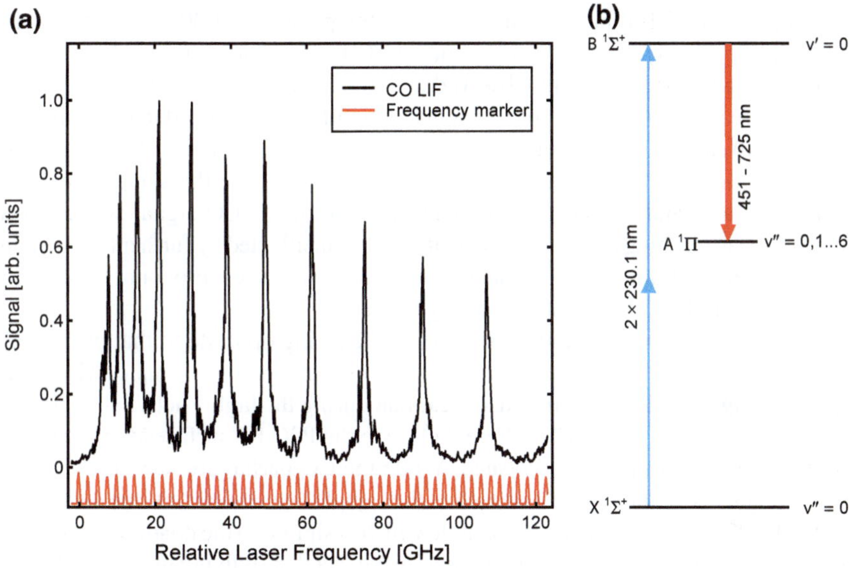

Fig. 6.10 **a** Excitation scan of CO gas at 50 mbar [2] (115 nm). **b** Schematic drawing of the two-photon excitation of the CO molecule and the subsequent decay by fluorescence at different wavelength [2]

660 nm, see Fig. 6.10b. Using a picosecond laser system consisting of a mode-locked Nd:YAG laser (PL2143C, Ekspla) with external amplifier (APL70–1100, Ekspla), these band states could be excited. We use the Nd:YAG third harmonic at 355 nm to pump an Optical Parametric Generator (PG 401-P80-SH, Ekspla), tuned to 230 nm. The pulse duration was 80 ps, the laser pulse repetition rate 10 Hz, and the line width at 230 nm was specified to 5 cm^{-1}. For CO PLIF imaging the 230 nm laser beam was directed through the reactor and shaped into a 7 mm high vertical sheet using two cylindrical lenses of focal lengths f = +500 mm and f = +300 mm. Images were acquired using an f = 50 mm objective (Nikkor f/1.2) and a 36 mm extension tube mounted on an ICCD camera (PIMAX III, Princeton Instruments). Images were acquired at 10 Hz repetition rate every 100 ms with the intensifier gate set to 30 ns. A longpass filter (GG395, Schott) was used to suppress scattering and fluorescence from surfaces in the reactor. The CO pressures were calibrated by using gas mixtures of well-defined CO partial pressures at a reactor temperature of 150 °C.

6.4.1 CO Imaging Above a Pd(110) Single Crystal

Figures 6.11a, b, and c show CO PLIF images of the gas phase above a Pd(110) single crystal in partial pressures of 26.5 mbar CO, 26.5 mbar O_2, and 53 mbar Ar, yielding a total pressure of 106 mbar in the reactor, at temperatures of 300 °C, 365 °C, and 375 °C, respectively. At 300 °C a significant CO fluorescence can be detected while a decreasing CO signal can be observed at 370 °C, indicating an onset of the CO_2 oxidation reaction. At 370 °C the CO signal is significantly reduced in the vicinity of the Pd(110) sample, suggesting that the surface is in a highly active state. Fig. 6.11d shows the intensity of the LIF signal in the area just above the sample indicated in Fig. 6.11a, b and c. Here a prominent decrease of the CO signal is confirmed above 370 °C reaching a low-level plateau which is unaffected by further temperature increase, demonstrating a MTL reaction rate. A similar, but less pronounced behavior can be seen in the CO mass spectrometry signal.

It is gratifying to see that the CO signal is behaving opposite to that of the CO_2 signal observed in the case of CO_2 imaging above the Pd(100) sample, which was shown in Fig. 6.6. However, it is also clear that during the highly active state the CO partial pressure is significantly different close to the Pd(110) surface as compared to at the outlet, where the MS is located. While the CO partial pressure decreases by 80% 0.3 mm above the Pd(110) surface as the sample becomes active, the decrease is only 20–30% 9 mm away from the sample which is similar to the condition where the MS is located. These observations emphasize the fact that studying model catalysts in a high-speed reaction such as CO oxidation in a vacuum chamber using semi-realistic conditions demands gas-phase measurements in direct vicinity of the catalyst.

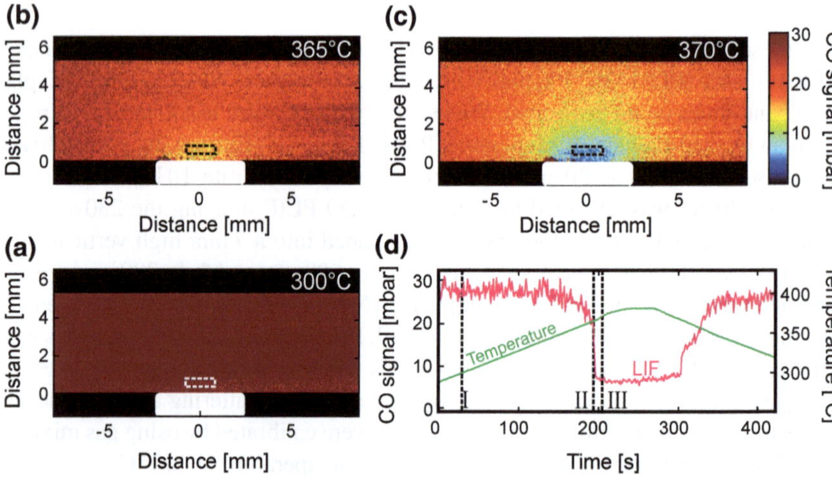

Fig. 6.11 CO LIF images (averages over ten single-shot frames) from a Pd(110) crystal in a gas composition of 26.5 mbar CO, 26.5 mbar O_2, and 53 mbar Ar at **a** 300 °C, **b** 365 °C, **c** 370 °C. **d** LIF intensity plot (*red*) from the indicated non-filled square in the LIF images and the temperature (*green*)

6.5 NH_3 Imaging

The potential to use PLIF to image the gas phase surrounding an active catalyst is significant. At present, only one gas species can be probed for each experiment. However, the experience from other research areas, such as combustion physics, in handling all aspects of laser techniques makes it possible to probe a large number of molecules important for catalysis. As an example, we present how NH_3 can be imaged by PLIF as it takes part in a catalytic reaction.

In the case of the NH_3 presented here, two-photon LIF was used as has been described elsewhere [37]. In Fig. 6.12a an excitation scan of the NH_3 molecules is shown. The simultaneous absorption of two photons allows for excitation wavelengths in the visible regime facilitating the detection capabilities. To this end, a combined Nd:YAG (Quanta-Ray PRO 250-10, Spectra Physics) and dye laser (Cobra Stretch-G-2400, Sirah) system operating at 10 Hz repetition rate and with 8 ns pulse duration was used for two-photon excitation of NH_3. The fluorescence emission spectrum is shown in Fig. 6.12b, showing best detection capabilities at wavelengths around 565 nm which for the present LIF measurements was detected with an intensified CCD camera (PI-MAX3, Princeton Instruments). A schematic of the two-photon excitation and the fluorescence emission is shown in Fig. 6.12c.

6.5.1 *NH_3 Oxidation Above a Ag/Al_2O_3 Powder Catalyst*

The investigation of the oxidation of NH_3 is part of a project to investigate the NO reduction reaction in an oxidizing environment using Ag-based catalysts [38] and NH_3 or hydrocarbons as a reducing agent. Since the NO reduction by NH_3 or a hydrocarbon, and the oxidation of NH_3 or the hydrocarbon by oxygen is competing, it is also of interest to study the NH_3 oxidation. In the present investigation, we used a Ag/Al_2O_3 powder catalyst to study the abundance of NH_3 in different envi-

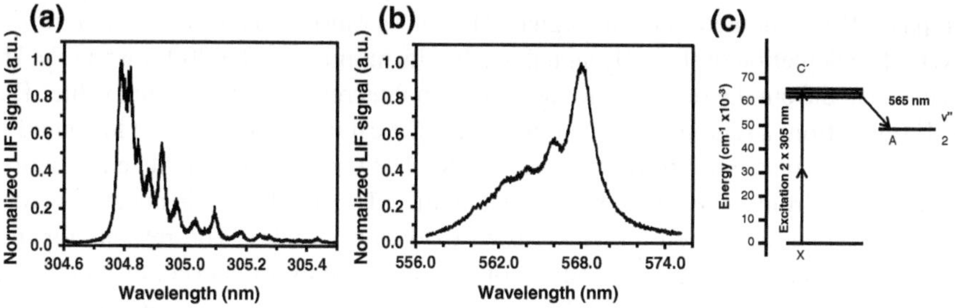

Fig. 6.12 **a** Excitation scan of the NH_3 molecules. **b** Fluorescence emission spectrum from NH_3 gas [37]. **c** Schematics of the two-photon excitation and fluorescence detection energies used in the present measurements

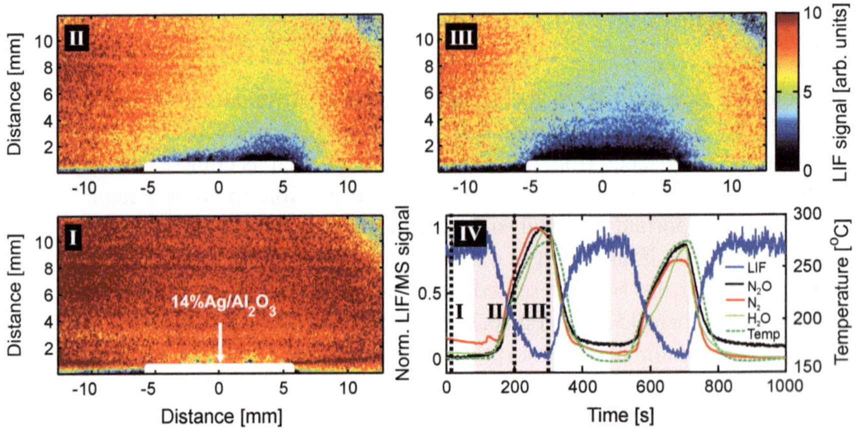

Fig. 6.13 NH_3 imaging (*panels I–III*) above a Ag/Al_2O_3 powder catalyst at a constant temperature of 160 °C and relevant mass spectrometry signals (*panel IV*). At 0 s a homogenous NH_3 distribution is observed. As soon as the H_2 is introduced (indicated by *red*), the NH_3 signal is depleted, reaching a minimum after 300 s, when the H_2 is evacuated

ronments. In particular, the experiments illustrate the promotional effect of H_2 on the oxidation of NH_3. The oxidation of NH_3 may follow several reaction pathways, for example

$$\mathbf{2NH_3 + 3/2O_2 \Rightarrow N_2 + 3H_2O}$$
$$\mathbf{2NH_3 + 5/2O_2 \Rightarrow 2NO + 3H_2O}$$
$$\mathbf{2NH_3 + 7/2O_2 \Rightarrow 2NO_2 + 3H_2O}$$

In Fig. 6.13 we show an experiment using a Ag/Al_2O_3 powder catalyst inside the reactor at a temperature of approximately 160 °C in a mixture of 3.6% NH_3, 90% O_2 with pulses of 6% Ar or H_2. The PLIF images are shown in panels I-III, and after 10 s of constant pressures the NH_3 distribution appears homogenous above the catalyst. However, as soon as H_2 is introduced into the reactor, the NH_3 LIF signal just above the catalyst starts to decrease in intensity and after 200 s it is clearly visible in panel II as a depletion of the signal. This observation indicates oxidation of the NH_3. The depletion of the NH_3 signal reaches its minimum after 300 s, when the H_2 is evacuated from the reactor, and the NH_3 signal increases to its original value. The LIF signal from a region of interest just above the catalyst is also shown in panel IV, as well as the mass spectrometry signals of N_2, H_2O, and N_2O, and the temperature.

It is clear from the PLIF NH_3 images that the NH_3 distribution is not even across the catalyst surface when the H_2-assisted oxidation initiates. The reasons for this could be either an uneven distribution of the active catalyst material, or it could be due to slight inhomogeneities in the temperature distribution of the sample.

6.6 Summary

PLIF in combination with the current fast development of laser and imaging technology has provided a new 2D-view on catalysis. Our initial driving force has been the urge to provide a 2D visualization of the CO_2 distribution around a single-crystal model catalyst during a CO oxidation experiment in semi-realistic conditions, to understand previous surface-science-based *in situ* results.

The present chapter demonstrates that it is currently possible, without too much effort, to measure the CO_2 signal during a CO oxidation experiment, providing that expertise in both laser diagnostics and catalysis-related research is combined. By choosing the right laser and detector combination, the 2D CO_2 distribution can be followed on a timescale faster than 0.4 s. This enables one to directly follow the ignition of the CO_2 production from single crystals or technical powder catalysts and the direct 2D visualization of the CO_2 boundary layer at MTL conditions, which demonstrates the substantial differences in gas composition close to the catalyst as compared to far away from the catalyst. The 2D-aspect of CO_2 PLIF imaging also enables a direct view of the CO_2 ignition of two samples in the same reactor, differentiating between their ignition temperatures. Furthermore, the chapter also provides indications on how spill-over effects can be avoided by introduction of exhaust tubes into the reactor.

Similarly, by the use of the correct laser and detection system, the 2D CO distribution in the same reaction can also be visualized and measured. Here, the intensity variation is inverted as compared to the CO_2 measurements, and directly explains observations made by previous high-pressure XPS measurements.

Finally, this chapter also describes how to perform PLIF measurements from NH_3 as it is oxidized in a catalytic oxidation reaction above a powder catalyst.

It is clear that PLIF applied to catalysis is not limited to the molecules probed in the present chapter. The extensive experience from the combustion physics/chemistry community will enable PLIF measurements of most small hydrocarbons as well as from a large range of other molecules as reactants or products in the vicinity of a catalyst for a particular reaction at work. Therefore, the potential benefits of applying PLIF to catalysis, for a better understanding of certain catalytic reactions, to identify intermediates only present in the gas phase in the vicinity of the catalyst, for potential commercial applications, and for combining with other experimental *in situ* techniques to simultaneously probe the presence of surface species is enormous.

Acknowledgements This work was financially supported by the foundation for strategic research (SSF), the Swedish Research Council, the Crafoord Foundation, the Knut and Alice Wallenberg Foundation, the Anna and Edwin Berger Foundation and NordForsk.

References

1. S. Svanberg, Phys. Scripta **T19b**, 469 (1987)
2. M. Aldén, S. Wallin, W. Wendt, Appl. Phys. B-Photo **33**, 205 (1984)
3. A.C. Eckbreth, *Laser Diagnostics for Combustion Temperature and Species, Abacus Press* (Kent; Cambridge, Mass, Tunbridge Wells, 1988)
4. E. Fridell, U. Westblom, M. Aldén, A. Rosén, J. Catal. **128**, 92 (1991)
5. E. Fridell, A.P. Elg, A. Rosén, B. Kasemo, J. Chem. Phys. **102**, 5827 (1995)
6. E. Fridell, A. Rosén, B. Kasemo, Langmuir **10**, 699 (1994)
7. F. Gudmundson, J.L. Persson, M. Forsth, F. Behrendt, B. Kasemo, A. Rosén, J. Catal. **179**, 420 (1998)
8. W. Kang, O. Fujita, K. Ito, J. Energ. Resour-Asme. **118**, 82 (1996)
9. F. Gudmundson, E. Fridell, A. Rosén, B. Kasemo, J. Phys. Chem-Us **97**, 12828 (1993)
10. H. Su, E.S. Yeung, J. Am. Chem. Soc. **122**, 7422 (2000)
11. B.L.M. Hendriksen, S.C. Bobaru, J.W.M. Frenken, Surf. Sci. **552**, 229 (2004)
12. R. van Rijn, O. Balmes, A. Resta, D. Wermeille, R. Westerstrm, J. Gustafson, R. Felici, E. Lundgren, J.W.M. Frenken, Phys. Chem. Chem. Phys. **13**, 13167 (2011)
13. J. Gustafson, R. Westerström, A. Mikkelsen, X. Torrelles, O. Balmes, N. Bovet, J.N. Andersen, C.J. Baddeley, E. Lundgren, Phys. Rev. B **78**, 045423 (2008)
14. R. Westerström, J.G. Wang, M. Ackermann, J. Gustafson, A. Resta, A. Mikkelsen, J.N. Andersen, E. Lundgren, O. Balmes, X. Torrelles, J.W.M. Frenken, B. Hammer, J. Phys.: Condens. Matter **20**, 184019 (2008)
15. F. Gao, S.M. McClure, Y. Cai, K.K. Gath, Y. Wang, M.S. Chen, Q.L. Guo, D.W. Goodman, Surf. Sci. **603**, 65 (2009)
16. F. Gao, Y. Wang, Y. Cai, D.W. Goodman, J. Phys. Chem. C **113**, 174 (2009)
17. J. Gustafson, R. Westerström, A. Resta, A. Mikkelsen, J.N. Andersen, O. Balmes, X. Torrelles, M. Schmid, P. Varga, B. Hammer, G. Kresse, C.J. Baddeley, E. Lundgren, Catal. Today **145**, 227 (2009)
18. J. Gustafson, R. Westerström, O. Balmes, A. Resta, R. van Rijn, X. Torrelles, C.T. Herbschleb, J.W.M. Frenken, E. Lundgren, J. Phys. Chem. C **114**, 4580 (2010)
19. J. Gustafson, R. Westerström, O. Balmes, A. Resta, X. van Rijn, R. Torrelles, C.T. Herbschleb, J.W.M. Frenken, E. Lundgren, J. Phys. Chem. C **114**, 22372 (2010)
20. R. Toyoshima, M. Yoshida, Y. Monya, Y. Kousa, K. Suzuki, H. Abe, B.S. Mun, K. Mase, K. Amemiya, H.J. Kondoh, Phys. Chem. C **116**, 18691 (2012)
21. S. Blomberg, M.J. Hoffmann, J. Gustafson, N.M. Martin, V.R. Fernandes, A. Borg, Z. Liu, R. Chang, S. Matera, K. Reuter, E. Lundgren, Phys. Rev. Lett. **110**, 117601 (2013)
22. R. Toyoshima, M. Yoshida, Y. Monya, K. Suzuki, K. Amemiya, K. Mase, B.S. Mun, H. Kondoh, J. Phys. Chem. C **117**, 20617 (2013)
23. J. Gustafson, S. Blomberg, N.M. Martin, V. Fernandes, A. Borg, Z. Liu, R. Chang, E. Lundgren, J. Phys.: Condens. Matter **26**, 055003 (2014)
24. J. Gustafson, M. Shipilin, C. Zhang, A. Stierle, U. Hejral, U. Ruett, O. Gutowski, P.A. Carlsson, M. Skoglundh, E. Lundgren, Science **343**, 758 (2014)
25. J. Rogal, K. Reuter, M. Scheffler, Phys. Rev. Lett. **98**, 046101 (2007)
26. J. Rogal, K. Reuter, M. Scheffler, Phys. Rev. B **77**, 155410 (2008)
27. S. Matera, K. Reuter, Catal. Lett. **133**, 156 (2009)
28. S. Matera, K. Reuter, Phys. Rev. B **82**, 035410 (2010)
29. Z. Duan, G. Henkelman, ACS Catal. **4**, 3435 (2014)
30. J. Zetterberg, S. Blomberg, J. Gustafson, Z.W. Sun, Z.S. Li, E. Lundgren, M. Aldén, Rev. Sci. Instrum. **83**, 053104 (2012)
31. B.J. Kirby, R.K. Hanson, Appl. Optics. **41**, 1190 (2002)
32. B.J. Kirby, R.K. Hanson, Appl. Optics. **40**, 6136 (2001)
33. Z.S. Li, M. Rupinski, J. Zetterberg, Z.T. Alwahabi, M. Aldén, Chem. Phys. Lett. **407**, 243 (2005)

34. B.J. Kirby, R.K. Hanson, Proc. Combust. Inst. **28**, 253 (2000)
35. J.T. Houghton, P. Phys, Soc. London **91**, 439 (1967)
36. W.A. Rosser, A.D. Wood, E.T. Gerry, J. Chem. Phys. **50**, 4996 (1969)
37. C. Brackmann, O. Hole, B. Zhou, Z.S. Li, M. Aldén. Appl. Phys. B **115**, 25 (2014)
38. F. Thibault-Starzyk, E. Seguin, S. Thomas, M. Daturi, H. Arnolds, D.A. King, Science **324**, 1048 (2009)

Chapter 7
Ab Initio Thermodynamics and First-Principles Microkinetics for Surface Catalysis

Karsten Reuter

Abstract Ab initio thermodynamics and first-principles microkinetic simulations have become standard tools in research on model catalysts. Complementing dedicated in situ experiments, these techniques contribute to our evolving mechanistic understanding, in particular of a reaction-induced dynamical evolution of the working catalyst surface. This topical review surveys the methodological foundations and ongoing developments of both techniques, and specifically illustrates the type of insights they provide in the context of in situ model catalyst studies. This insight points at substantial deviations from the standard picture that analyzes catalytic function merely in terms of properties of and processes at active sites as they emerge from a crystal lattice truncation of the nominal catalyst bulk material.

7.1 Introduction

An obvious target of research in heterogeneous catalysis is to develop "better" catalysts. "Better" may thereby stand for quite different aspects. Among others this can be higher activity, higher selectivity, longer lifetimes, or preferable materials. Whatever the targeted improvements are specifically though, if they are to be found by anything but mindless trial and error, one necessarily needs "ideas". One powerful source of ideas to find better catalysts is to understand what limits the function of existing catalysts. Generally, the better or detailed this understanding is, the better defined are the ideas that emerge from it. This line of thinking is the basic motivation for catalysis research that aims for what one refers to as mechanistic understanding. Here, mechanistic ideally means understanding the function down to the atomistic level of the individual elementary processes that underlie the catalytic

This chapter has also been published as Catalysis Letters 146 (3), 541—563 (2016).

K. Reuter (✉)
Chair for Theoretical Chemistry and Catalysis Research Center, Technical University Munich, Lichtenbergstr. 4, 85747 Garching, Germany
e-mail: karsten.reuter@ch.tum.de

J. Frenken and I. Groot (eds.), Operando *Research in Heterogeneous Catalysis*,
Springer Series in Chemical Physics 114, DOI 10.1007/978-3-319-44439-0_7

cycle. It turns out that this is a pretty daunting goal. One possibility to make it at least a bit more tractable is to reduce the complexity of the problem and achieve this understanding first for model catalysts [1, 2], i.e. typically single-crystals of the actual catalyst material or defined nanoparticles of the material on single-crystal supports. This dismisses many aspects of a real catalyst, and may therefore only generate a subset of ideas—but, one has to make a start.

One of the central, novel aspects that have recently emerged from such mechanistic studies on model catalysts is that an operating catalyst surface could be anything but a static entity [3, 4]. A prevailing view of heterogeneous catalysis often found in introductory textbooks is instead that of impinging and reacting gas-phase species on a rigid solid surface [5–7]. If the atomic structure of the surface is resolved at all in such a picture, then this is the crystallographic structure as resulting from a mere surface truncation of the bulk catalyst lattice. For instance, for metal catalysts one pictures a low-index facet like a (111) or (100) fcc surface, flat like a tablet, at best with some steps in between. The surface metal atoms have a reduced metal coordination in comparison to the coordination of a bulk atom. This makes them "active" and one views particular high-symmetry adsorption sites on the lattice defined by the position of these "active" surface atoms as the ones driving the catalysis. Consequently denoted as "active sites", in the example of the fcc metal surface this could e.g. be hollow, bridge or atop adsorption sites on the terraces, or equivalent sites at upper or lower step edges. The surface metal atoms around these active sites adapt their positions slightly to the ongoing elementary processes of the catalytic reaction, namely adsorption, diffusion, reaction, and desorption of the reactants and reaction intermediates. However, apart from such small structural relaxations, the surface morphology is assumed to be pretty static. As such, the catalytic function is analyzed in terms of the properties of and processes at these active sites, thinking specifically of sites as they emerge from the crystal lattice truncation of the nominal catalyst bulk material.

While seductively familiar and intuitive, this picture could fall short in capturing much of real heterogeneous catalysis. For sure, the picture is largely correct in the defined environment offered by controlled gas dosage in ultra-high vacuum (UHV) and at low temperatures. Most of what we know on an atomic level about surface catalytic reactions derives from such environments and this is why the above sketched picture is familiar and intuitive to us. However, heterogeneous catalysis does not operate in UHV. Technologically relevant gas-phase conditions comprise ambient pressures or beyond. Under a corresponding, much fiercer gas-phase impingement we at least have to expect increased adsorbate concentrations at the surface and concomitantly higher reaction rates, typically measured in turnover frequencies (TOF) with units of product molecules per catalyst surface area and time. If this was all, it should still be possible to extrapolate from UHV to ambient conditions and to slow things down by studying lower temperatures. Such "thermodynamic scaling" (vide infra) was the original hope or assumption of the Surface Science approach to Heterogeneous Catalysis. Increasingly, we are able to scrutinize this assumption. This is made possible by the advent of so-called in situ studies that investigate model catalysts at ideally similar atomic resolution as in traditional UHV Surface Science, but at (near-)ambient pressures [8, 9]. What we have learnt from such studies so far, sketches a picture of heterogeneous catalysis that goes far beyond a simple thermodynamic scaling.

For instance, the surface concentrations of certain reaction intermediates can become so high that phase transitions to new compound materials composed of the original (nominal) catalyst material and the reaction intermediate may occur—and it is this new material that then actuates the catalysis [4, 8, 9]. Both for thermodynamic or kinetic reasons these new materials must furthermore by no means be restricted to known bulk phases. Instead they can exhibit completely new structures that are (temporarily) stabilized for instance as thin surface films on top of the bulk catalyst. A prominent example for such surface morphological transitions is oxide formation at the late transition metals employed in oxidation catalysis [10–14]. Another aspect that speaks against a simple scaling from UHV to ambient pressures is the much higher amount of reaction energy that is released in case of exothermic reactions at the increased reaction rates. We presently know very little about how and how quickly this energy is dissipated on an atomic-scale [15]. Yet, if heat transfer is limited, scenarios like molten catalyst materials with a surface dynamics much beyond that of rigidly lattice-aligned active sites are well conceivable.

One needs to stress that the current understanding we have gained through in situ studies is far from being complete; Certainly much less than what we have collected in decades of UHV Surface Science work. At present it is not clear whether those instances reported are exotic oddities or the top of the iceberg. The data we already have is nevertheless good enough to formulate a working hypothesis opposite to the prevalent static picture: Why not view a catalyst surface as something entirely dynamic? A surface that while operating adapts sensitively to the reaction conditions in everything ranging from the local atomic structure to overall composition and morphology? Yes, new surface phases can form in the reactive gas-phase, but why should they always cover the entire catalyst surface? As a result, if the surface is then heterogeneous, why should this surface heterogeneity not vary with time? Maybe new active site configurations form and decay continuously as a result of interaction with the reactants and reaction intermediates, and maybe they even form specifically at phase boundaries arising on the evolving surface. Clearly, the only possibility we have to validate or falsify such "ideas"—and the consequences they would suggest for the design of "better" catalysts—is to study the catalyst not in UHV, not before it goes on stream, not after it has gone out of stream, but precisely *operando*, when it is working under technologically relevant gas-phase conditions.

This has exactly been the motivation of the aforementioned in situ studies on model catalysts that have made their fulminant appearance over the last decade or so [8, 9]. At spatial and temporal resolution that is ever increasing and at pressures that come closer and closer to technological conditions, such studies precisely focus on the surface structure, composition and morphology—and try to relate it to the catalytic activity. Aiming e.g. to extend the use of UHV electron spectroscopies to these pressure regimes, the experimental setups are necessarily involved. Mass flow limitations in the resulting complex reactor chambers together with still limited resolving powers render the measured data not always straightforward to interpret

[16–18]. As in many areas of materials and chemical science much synergy has therefore been achieved by complementing these measurements with modern computational theory. To a large degree these are the same (typically first-principles based) calculations of thermostability, spectroscopic signals, and reactivity descriptors as they have already been successfully conducted for a long time in the realm of UHV Surface Science. In addition, however, new theoretical approaches have been developed and advanced that have exactly the same objectives as the in situ studies [19]: For given reaction conditions in form of defined reactant partial pressures p_i and temperature T, what is the surface structure and composition—and what is the corresponding catalytic activity?

Aiming to provide this information independently, i.e. be of predictive character, such theory necessarily has to be based on first-principles electronic structure calculations. In order to account for the effect of finite temperature and pressure, as well as for the ensemble character introduced by the ongoing surface chemical reactions, these quantum mechanical calculations need to be combined with concepts from thermodynamics and statistical mechanics. Notably, two such approaches have been established that have proven so powerful that they are nowadays firmly anchored in the conceptual toolbox of everybody working in surface catalysis: (constrained) ab initio thermodynamics and first-principles microkinetics. The prior technique provides exclusively access to the surface structure and composition as a function of (T, p_i). The theory is approximate, but therefore computationally less intense and applicable to more complex surface structures. As the name implies, first-principles microkinetics explicitly accounts for the kinetic effects due to the ongoing chemical reactions. It is therefore intrinsically more accurate and additionally gives access to the catalytic activity. This comes at the price of larger computational cost and, at least in its most rigorous implementations, presently still with quite some restrictions with respect to the complexity of the surface structures and reaction networks it can handle. In this topical review I will survey both techniques, yet not so much in terms of their detailed methodological foundations and technical implementations. Extended reviews are available for this [19–22]. Instead, I will focus on their concepts, discuss some current frontiers and ongoing developments, and specifically illustrate the type of insights they provide in the context of in situ model catalyst studies.

7.2 (Constrained) Ab Initio Thermodynamics

7.2.1 Methodology

Even though they form the basis of both techniques that will be covered I will not at all dwell on the underlying first-principles electronic structure calculations [23]. In the context of in situ studies on model catalysts these calculations are at present almost exclusively performed within density-functional theory (DFT). The central output of these calculations that enters into the first-principles thermodynamics or statistical mechanics approaches is the total energy E^{tot}, i.e. the energy contained in the chemical bonds for a defined structural configuration of atoms. For the present purposes there are two aspects of these total energies that need to be highlighted: First, the total energies are only approximate, which is primarily due to the approximate exchange-correlation (xc) functional that is employed in the DFT calculations [24]. In fact, due to the typically rather large system sizes, computationally less demanding, so-called lower-rung xc functionals are predominantly applied [19, 25, 26]. For metal catalysts these are largely still semi-local generalized gradient approximation (GGA) functionals, while for materials with more localized bonding aspects like oxides these are increasingly hybrid functionals. What this implies is that we have to expect an uncertainty in central quantities like binding energies (suitable differences of total energies) or reaction barriers (difference of binding energies at initial and transition state) that is of the order of ~0.3 eV (~30 kJ/mol). Of course, since we lack the exact xc functional this is only a rough estimate, and for reaction barriers some error cancelation when taking a difference from differences might make the uncertainty a bit smaller. Notwithstanding, the latter is more a hope than something to rely on. In any case, we thus have to count with potential errors that are much larger than k_BT. This obviously has to be kept in mind when attempting to make statements about temperature-dependent properties or even more so about catalytic activities where reaction barriers enter through exponential Boltzmann-type factors.

The other aspect to highlight is to repeat that $E^{tot} = E^{tot}(N_i, N_j)$, where N_i and N_j are the number of species i and j in the particular configuration that has been calculated. I distinguish here and in the following between species i that are also present in the gas phase (i.e. contained in the reactants), and species j that are not (i.e. that are only present on the solid catalyst). A straightforward comparison of the stability of two configurations on the basis of DFT total energies is therefore only possible, if both configurations contain exactly the same numbers $N'_i = N_i$ and $N'_j = N_j$ of all species i and j in the system. On the contrary, in the context of near-ambient catalysis the surface composition is precisely one of the targeted quantities, i.e. one does a priori not know how many atoms of which kind there are in the surface fringe. As already pointed out before, to the very least one would expect surface coverages of reaction intermediates to change with varying pressures. In order to find out which coverage there is for given reactant partial pressures p_i, one would thus have to compare the stability of configurations with different coverages, i.e. with differing numbers N_i. This is precisely what cannot be achieved on the basis of the E^{tot} alone. In order to answer such questions one would need to know the cost of bringing the difference in species $\Delta N_i = N'_i - N_i$ and $\Delta N_j = N'_j - N_j$ between two configurations either into one of the configurations or out of the other configuration. Thermodynamically, it would thereby not matter through which particular (atomistic) mechanism this happens. The only thing that would matter is where they ultimately come from or go to.

The entire idea of ab initio thermodynamics is to provide this information by considering such reservoirs where species go to or come from, and then work within an appropriate thermodynamic framework to compare configurations with varying numbers of species [27–35]. Since experiments and heterogeneous catalysis are generally run at defined temperatures and reactant partial pressures, the appropriate thermodynamic ensemble for this framework is the Gibbs ensemble (T, p_i). To assess the stability of a given surface configuration, a suitable quantity to evaluate is then for instance the surface free energy per surface area A,

$$\gamma(T,p_i) = \frac{1}{A}\left[G(T,p_i,N_i,N_j) - N_i\mu_i - N_j\mu_j\right], \tag{7.1}$$

where G is the Gibbs free energy of a particular surface configuration containing N_i species i and N_j species j, and μ_i and μ_j are the chemical potentials of the corresponding reservoirs of species i and j. This surface free energy represents the cost of creating the particular surface configuration by taking all of its constituent atoms out of their respective reservoirs. Calculating $\gamma(T, p_i)$ for a range of potential surface configurations, the one that exhibits the lowest surface free energy is this way the most stable one that (if thermodynamics is correct) should be observed in experiment.

In order to evaluate (7.1) for a given surface configuration one needs to know the chemical potentials. For any gas-phase species i, the obvious reservoir that determines this chemical potential is the gas-phase environment itself. Approximating this gas phase as an ideal gas, it is straightforward to obtain $\Delta\mu_i = \Delta\mu_i(T, p_i)$, where $\Delta\mu_i = \mu_i - E^{tot}(i)$ and $E^{tot}(i)$ is the DFT total energy of the isolated gas-phase species i. For atoms and small molecules this can even be calculated analytically [36, 37]. For others, values can be found in thermodynamic tables [38]. For the other species j that are not present in the gas phase, e.g. species constituting the actual catalyst material, alternative reservoirs need to be found. This can often be facilitated by choosing a suitable reference configuration and evaluating only the excess surface free energy with respect to this reference

$$\gamma(T,p_i) - \gamma_o(T,p_i) = \frac{1}{A}\left[G(T,p_i,N_i,N_j) - G_o\left(T,p_i,N_i',N_j'\right) - \Delta N_i\mu_i - \Delta N_j\mu_j\right], \tag{7.2}$$

where γ_o and G_o are the surface free energy and Gibbs free energy of the reference configuration, respectively. This has the advantage that one only needs to define reservoirs for non-gas-phase species j, for which $\Delta N_j \neq 0$. If we are for instance interested in evaluating the relative stability of different surface coverages of a given reactant on the surface of a solid catalyst, then the clean surface at zero reactant coverage is obviously a useful reference configuration. In this case we would only need to determine a suitable reservoir for species constituting the catalyst and not being present in the gas phase, if the changing concentration of adsorbed reactants would actually affect the density of catalyst species in the surface fringe. In the case of compound materials, such a density change could thereby for example proceed via precipitation of another (bulk) phase. In this case, this other phase, say for instance pure metal droplets at the surface of metal oxide catalysts, is in turn a suitable reservoir defining the chemical potential of the substrate species involved in the density change.

What is thus left to get hard numbers out of (7.1) or (7.2) is to compute the solid-state Gibbs free energies (and equivalently the chemical potentials of the non-gas-phase reservoirs). Similar to the procedure for the gas-phase species, it is thereby useful to separate off the DFT total energy, which in terms of thermodynamic potentials amounts to the Helmholtz free energy minus the zero point energies. We thus have $G = E^{tot} + \Delta G(T, p_i)$, where the zero point energies are now considered to be contained in the temperature and pressure dependent free energy part $\Delta G(T, p_i)$. In comparison to the ideal gas situation, calculating this free energy part is more involved for solids, and, unfortunately, the term itself is also generally not negligible [39–41]. Fortunately, however, it is not this absolute free energy part that matters for calculating in particular the excess surface free energy. As apparent from (7.2) it is only the difference of two solid-state Gibbs free energies and additional chemical potentials that enters, and in this difference many contributions can cancel. Since the predominant contribution to solid-state $\Delta G(T, p_i)$ comes from vibrational free energy, it is thus not the absolute vibrations that enter. Instead it is only changes of these vibrations (phonon spectrum, to be precise) with respect to the reference configuration that matter, and these changes can often be neglected for a first assessment. In the difference of (7.2)—and only there—we can then write

$$\gamma(T,p_i) - \gamma_o(T,p_i) \approx \frac{1}{A}\left[\Delta E^{tot} - \Delta N_i \Delta\mu_i(T,p_i)\right] \tag{7.3}$$

with

$$\Delta E^{tot} = E^{tot}(N_i, N_j) - E_o^{tot}\left(N_i', N_j'\right) - \Delta N_i E^{tot}(i) - \Delta N_j E^{tot}(j)$$

and $E^{tot}(j)$ the DFT total energy of the solid-state reservoir chosen for species j. In this approximation, the computational demand to evaluate the excess surface free energy of a given configuration is therefore reduced to DFT calculations of the surface configuration, the reference surface configuration, as well as of all isolated gas-phase species and the chosen additional solid-state reservoirs. The entire temperature and pressure dependence is instead exclusively contained in the terms $\Delta N_i \Delta\mu_i(T,p_i)$, where $\Delta\mu_i(T,p_i)$ is a look-up quantity that is generic for the species and not for the particular systems studied.

It is this low computational demand that makes this formulation of ab initio thermodynamics so appealing. One has to stress that this holds only within this prevalent approximation though. The neglected terms $\Delta G(T, p_i)$ are more involved. This refers thereby less to the predominant vibrational contribution to these terms, which can be and needs to be at least approximately calculated in many cases [39, 41]. The more elusive contribution comes instead from the configurational entropy [20]. Fortunately, for not too high temperatures this entropy is not large and for hitherto typically studied systems neglecting it affects the results only in a predictable way that I will further discuss below.

There is, however, a much more critical aspect than the neglect of these individual solid-state free-energy contributions. This is the way how the configurational sampling is performed in present applications of this theory. What is generally done is to consider a given number of configurations that is presumed to be of relevance for the problem at hand. Computing the excess surface free energies for every one of them, one of these configurations will exhibit the lowest value for given gas-phase conditions (T, p_i). This configuration is then declared to be most stable for these conditions, but obviously this statement can only refer to relative stability within the group of configurations that has actually been tested. If a configuration that was not considered was to exhibit an even lower excess surface free energy, it would not be identified. There is also no warning mechanism of such cases in any form in the present formulation of ab initio thermodynamics: The results obtained would simply be wrong. This limitation with respect to the considered configurations must always be born in mind when assessing the results of present-day ab initio thermodynamics studies. Of course, this limitation is not conceptual, but results merely from the steeply increasing computational costs when comparing extended numbers of configurations (certainly in the context of in situ studies of model catalysts). Any form of more systematic configurational sampling, as e.g. resulting from global geometry optimization algorithms, can be straightforwardly incorporated into the ab initio thermodynamics framework. The resulting total energies of all configurations sampled just need to be entered into thermodynamic equations of the type of equations (7.1)–(7.3), or one directly performs the sampling in the appropriate thermodynamic ensemble by evaluating different cost functions than the total energy.

7.2.2 Oxide Formation at (Near-)Ambient Conditions

After this brief methodological survey, let me illustrate the kind of insights and the effect of the discussed approximations and limitations with an application example in the in situ context. As mentioned before, a possible formation of oxides at the surface of late transition metal oxidation catalysts is a prototypical manifestation of the type of surface morphological transitions that one suspects to occur under technologically relevant, (near-)ambient reaction conditions. While nominally Rh, Pd, or Pt would thus be materials that one cites as catalysts employed for such reactions, in fact their oxides or "oxidic" films could be the ones that really actuate the catalysis. If true, it would obviously not make much sense to discuss the catalytic activity (and any "ideas" for improved catalysts) in terms of the classic active sites offered by fcc(111) of fcc(100) facets of these metals. One would have simply looked at the wrong material. Ab initio thermodynamics has been heavily employed in this context and a natural starting point is to only consider the effect of an oxygen environment. Using the clean metal surface as a suitable reference, one would calculate the excess surface free energy for a range of surface configurations with increasing oxygen content, and in turn evaluate their relative stabilities as a function of the oxygen chemical potential of the surrounding gas phase. Figure 7.1 exemplifies this for a Pd(100) surface [42–44]. Natural surface configurations to consider in such a case are various (ordered) simple adsorption layers of different concentrations as they could for example have been characterized in UHV Surface Science work. For the O/Pd(100) surface this would be so-called $p(2 \times 2)$ and $c(2 \times 2)$ structures with O atoms adsorbed at the fcc(100) hollow sites at ¼ monolayer (ML) and ½ ML coverage, respectively [45, 46]. The excess surface free energies of these structures will vary with varying O content in the gas phase. In (7.3) this enters through the linear dependence on the (oxygen) chemical potential, which scales differently for configurations with different amounts of (oxygen) species incorporated into the surface fringe. In the limit of an infinitely dilute gas ($\Delta\mu_O \rightarrow -\infty$), incorporating any O into the surface configuration would incur an infinite cost due to the concomitant infinite loss of entropy. This is why the clean surface reference naturally exhibits the lowest excess surface free energy under these conditions, cf. Fig. 7.1a.

With increasing oxygen content in the gas phase, $\Delta\mu_O$ will become less negative and it will become increasingly more favorable to stabilize oxygen at the surface. In

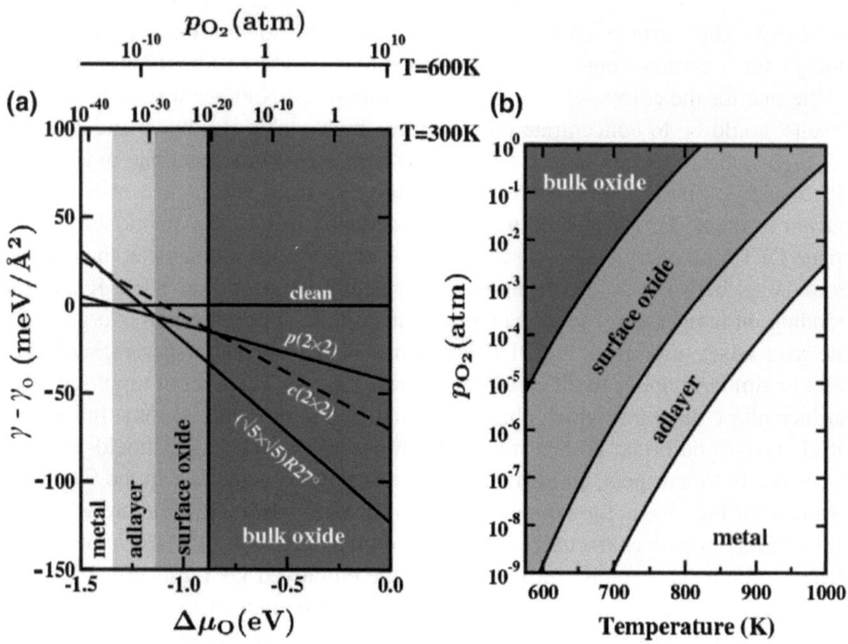

Fig. 7.1 **a** Excess surface free energies and **b** surface phase diagram for O/Pd(100). Considered are two ordered O adsorbate layers with different coverage ($p(2 \times 2)$, ¼ monolayer (ML), and $c(2 \times 2)$, ½ ML) and a $(\sqrt{5} \times \sqrt{5})$R27°–O surface oxide film (0.8 ML). Note the extended stability range of the surface oxide compared to the known PdO bulk oxide. The total energies (DFT-GGA, PBE) used to construct this graph via (7.3) are taken from [42–44]

the example of O/Pd(100) in Fig. 7.1a this happens at $\Delta\mu_O = -1.3$ eV, which is when the excess surface free energy of the $p(2 \times 2)$ adsorption structure becomes lower than the clean surface reference. The higher the surface O concentration of a configuration, the steeper will be the decrease of its excess surface free energy in a plot like Fig. 7.1a. This can eventually stabilize such configurations at higher O chemical potentials. The obvious upper limit of surface O concentration is thereby a complete transformation of the bulk metal into a bulk oxide, as this then implies an infinite number of O atoms in the normalization per surface area employed in (7.3) [47]. In a plot like Fig. 7.1a this leads to an infinite negative slope, i.e. a vertical excess surface free energy line. For the shown example of O/Pd(100) this line indicating the formation of bulk PdO lies at $\Delta\mu_O = -0.7$ eV, and for any higher oxygen chemical potential the PdO bulk oxide will be the stable phase.

Already at this stage it is worthwhile to point out what has been gained through this theory. On the basis of only a handful of static DFT calculations we can discuss the possible surface structure and composition at finite temperature and pressure. In a plot like Fig. 7.1a the latter two-dimensional (T, p_{O2})-dependence is thereby conveniently described through the one-dimensional dependence on the corresponding chemical potential. By defining suitable references one can convert one dependence into the other on an absolute scale [34]. As done in Fig. 7.1a this allows to include additional x-axes that give the pressure dependence at some specific temperature (or alternatively the temperature dependence at some fixed pressure). The surface configuration exhibiting the lowest excess surface free energy for a certain range of chemical potentials would be identified as the most stable one for the corresponding gas-phase conditions. Another way of plotting the results would be to concentrate only on these most stable structures and plot their (T, p_{O2})-stability ranges in a so-called surface phase diagram as done in Fig. 7.1b. Such surface phase diagrams are more intuitive to read, but there is also a certain *caveat* to them. This has to do with the uncertainty due to the mentioned approximate DFT total energies. For a surface phase diagram this implies that the obtained boundaries between different phases can typically be wrong by ~100 K and (depending on temperature) several orders of magnitude in pressure. As a large part of the error arises often from the DFT description of the gas-phase species, such shifts tend to similarly apply to all phase boundaries though. The overall topology of the surface phase diagram (which phases are predicted to be stable at some finite range of (T, p_{O2})-conditions) is then more robust, and this is what one should generally focus on. In this respect, an intriguing immediate result contained in the O/Pd(100) example of Fig. 7.1 is for example that the $c(2 \times 2)$ adsorbate structure which has been observed and characterized after gas dosage in UHV [45, 46] is never predicted to be a stable phase on the basis of the employed DFT functional.

A second intriguing aspect of ab initio thermodynamics that can be highlighted with the example of Fig. 7.1 is the possibility to test the stability of surface configurations one suspects to potentially play a role at finite temperatures and pressures. In the context of oxide formation this would prominently be thin oxide films at the surface. For O/Pd(100) such a structure had again be stabilized after excessive O dosage in UHV and was subsequently characterized as a layer of PdO(101) in a commensurate ($\sqrt{5}$ x $\sqrt{5}$)R27° stacking on top of Pd(100) [48, 49]. Evaluating the excess surface free energy for this surface structure, there is indeed a finite range of O chemical potentials where it is predicted to be most stable, cf. Figure 7.1. This range extends over $\Delta\mu_O$ lower than the ones of the known bulk oxide phase, i.e. ab initio thermodynamics predicts a range of less O-rich gas-phase conditions where bulk PdO is not yet stable, but such a PdO(101) overlayer is. Such an extended stability range of surface oxide films has been found for many low-index late transition metal facets [48–57]. It can arise from an enhanced coupling of the film to the underlying metal [58], but also simply because the structure of the thin films is by no means restricted to those of the known bulk oxides. The latter point thereby hints at the mentioned limitation of prevalent ab initio thermodynamics with respect to the configurational sampling. Maybe there are more complex, highly O-enriched surface configurations that would exhibit even lower excess surface free energies. Without knowing their explicit structure (or being able to represent this structure in computationally tractable periodic supercell geometries) their excess surface free energies cannot be calculated and the corresponding stabilities not be assessed. Even within the drive towards (near-)ambient catalysis this underscores the value of dedicated UHV Surface Science work that aims to stabilize and characterize such structures and therewith serves as a generator for structural models to test. Just as much as one might rather focus more on the overall topology of surface phase diagrams than their absolute phase boundaries, this also suggests that the really valuable "idea" that has emerged out of studies of the kind of the discussed O/Pd(100) work is not necessarily that of a particular, defined surface oxide structure. These ordered structures are likely just idealized models. Instead it is the general notion that such kind of O-enriched surface configurations (be they called surface oxides, oxidic films or sub-surface oxygen) can be stabilized in environments far less O-rich than those where bulk oxides are known to be stable.

7.2.3 Constrained Thermodynamics: Approximate Structure and Composition Under Reaction Conditions

Whether such configurations really play a role for (near-)ambient oxidation catalysis, then critically depends on the particular reaction. The presence of the other reactant tends to reduce the catalyst surface. In order to assess whether an oxidized configuration will prevail under reactive conditions, the other reactant thus needs to be accounted for. In ab initio thermodynamics this seems straightforward to do as a multi-component gas phase can simply be considered through multiple reservoirs for the corresponding gas-phase species [39]. In (7.1)–(7.3) this is already indicated through the dependence on several chemical potentials μ_i. There is a slight catch to this for heterogeneous catalysis though. If one was to consider full thermodynamic equilibrium, then also these various reservoirs would be in equilibrium with each other. However, if all reactants were in full equilibrium with each other, the gas phase would only consist of products, as a catalyst can only operate under gas-phase conditions where the products are thermodynamically more favorable than the reactants. As this is obviously not the situation we want to describe, one instead suitably resorts to a kind of "constrained" equilibrium approach [37, 59]. In order to capture the effect of exposure to the reactant gas phase, the catalyst surface is considered to be in full equilibrium with all reactant gas-phase chemical potentials, while the latter are treated as mutually independent of each other. The approximation that is introduced through this is to neglect that the actual on-going surface catalytic reactions may consume surface reaction intermediates faster than they can be replenished from the gas phase [60]. A "constrained" ab initio thermodynamics study can therefore only provide some first rough insight into the surface structure and composition in reactive environments, but its advantage is that, as before, a wide range of structurally and compositionally largely differing configurations can readily be compared in a computationally undemanding way.

Figure 7.2 illustrates this for the CO oxidation at Pd(100) system, where in contrast to Fig. 7.1 the CO chemical potential is now explicitly considered as a second axis [43, 44]. Comparing the stability of a large set of on-surface (co) adsorption, surface oxide and bulk oxide structures, several phases involving the $(\sqrt{5} \times \sqrt{5})R27°$ surface oxide are found to be most stable over a wide range of (T, p_{O2}, p_{CO})-conditions. Again, this range largely exceeds the stability range of bulk PdO. Intriguingly, this range extends in fact so much that it even just touches the gas-phase conditions typical for technological CO oxidation, i.e., partial pressures of the order of 1 atm and temperatures around 300–600 K. In terms of a potential oxide formation under reaction conditions, this would suggest that instead of thick bulk-like oxide films it would rather be such a nanometer-thin oxidic overlayer that could play a role. Indeed, in situ reactor scanning tunneling microscopy (STM) experiments observed substantial morphology changes that were precisely assigned to the formation of a thin oxidic overlayer [56, 61, 62]. However, in these experiments, a continuous consumption and formation of this surface oxide even under the employed steady-state reaction conditions was reported—which would directly relate to the general "idea" of a working catalyst as a very dynamic entity. For this aspect the proximity of the technologically-relevant (near-)ambient reaction conditions to the phase boundary between the surface oxide and reduced metal configurations in Fig. 7.2 has to be emphasized. In Fig. 7.2 this boundary is drawn as an infinitely sharp transition, whereas in reality any such phase transition would occur over a finite range of pressures and/or temperatures. This abrupt change in (μ_O, μ_{CO})-space in Fig. 7.2 is the result of the neglect of the solid-state configurational entropy contributions in (7.3). While these contributions are generally small compared to absolute excess surface free energies, they particularly matter for chemical potential conditions where the excess surface free energy lines of two competing configurations cross, i.e. exactly at phase boundaries. Under such conditions the thermally induced possibility to explore both configurations leads to enhanced fluctuations and phase coexistence [37, 59].

Under the neglect of configurational entropic contributions the prevalent formulation of (constrained) ab initio thermodynamics cannot explicitly account for such a phase coexistence (and the implied fluctuations). As done in Fig. 7.3 one

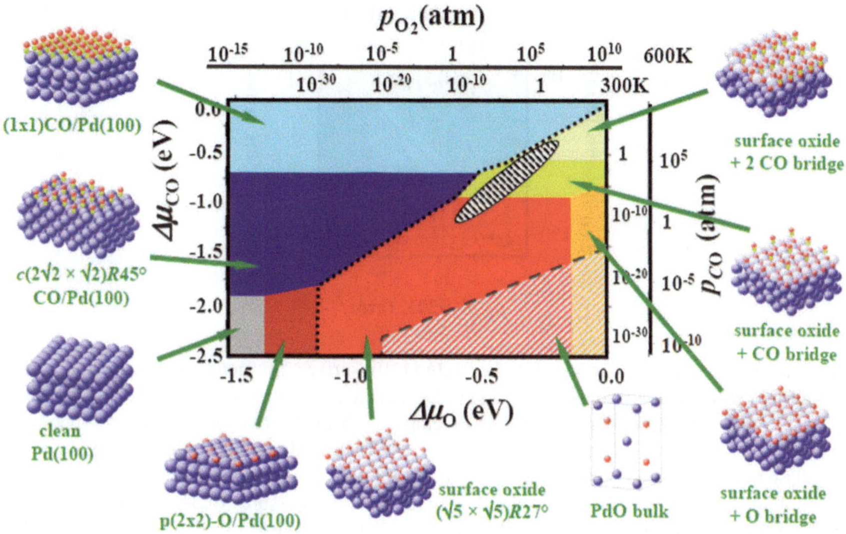

Fig. 7.2 Surface phase diagram for the Pd(100) surface in "constrained" thermodynamic equilibrium with an environment consisting of O_2 and CO. The atomic structures underlying the various stable (co-)adsorption phases on Pd(100) and the ($\sqrt{5} \times \sqrt{5}$)R27° surface oxide, as well as a thick bulk-like oxide film (indicated by the bulk unit-cell), are also shown (Pd: large blue spheres, O: small red spheres, C: white spheres). Phases involving surface or bulk oxide are to the right bottom of the dotted and dashed line, respectively. The dependence on the chemical potentials of O_2 and CO in the gas phase is translated into pressure scales at 300 and 600 K. The black hatched ellipse marks gas-phase conditions representative of technological CO oxidation catalysis, i.e., partial pressures of 1 atm and temperatures between 300 and 600 K. Adapted from [43]

may estimate its width in (T, p_i)-space and represent this information by drawing the phase boundaries as regions with a corresponding finite width [37, 59]. Figure 7.3 shows results equivalent to Fig. 7.2, but obtained for CO oxidation at RuO_2(110) [60]. Strikingly, technologically relevant feed conditions fall again precisely into such a phase coexistence region. The thus suggested notion to view heterogeneous catalysis as a phase transition phenomenon may thereby be rationalized by recalling that a so-called stable phase is not stable on an atomistic scale. Instead it represents an average over many continuously on-going processes such as dissociation, adsorption, diffusion, association, and desorption. As all these elementary processes and their interplay are of crucial importance for catalysis, regions in (T, p_i)-space that exhibit enhanced thermal fluctuations, i.e. where the dynamics of these atomistic processes is particularly strong, appear naturally as most relevant [37]. In this understanding where in phase space catalytically relevant regions might emerge, insights of the type provided by Figs. 7.2 and 7.3 also allow to comment on the possibility to further explore them by bridging the pressure gap between (near-)ambient real catalysis and UHV Surface Science. In the thermodynamic Gibbs ensemble the only ruling quantities are the chemical potentials μ_i.

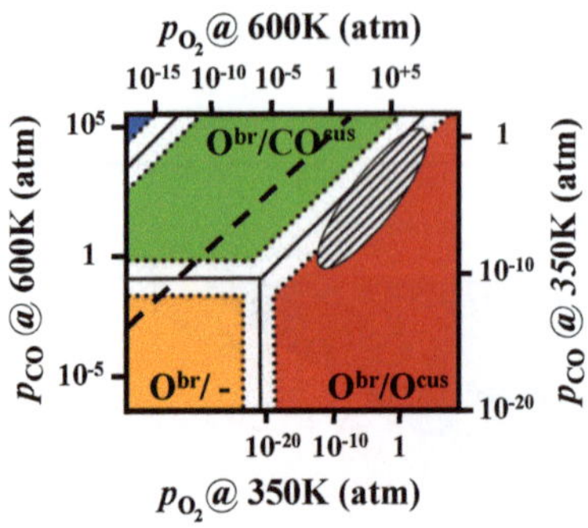

Fig. 7.3 Surface phase diagram for the $RuO_2(110)$ surface in "constrained" thermodynamic equilibrium with an environment consisting of O_2 and CO. The labels of the different stable phases reflect a predominant population (O, CO or empty "–") of the two prominent adsorption sites offered by this surface, br(idge) and coordinatively unsaturated (cus) site. Coexistence regions at the phase boundaries are marked in *white*, with the width of these regions corresponding to 600 K. Technologically relevant catalytic conditions around partial pressures of 1 atm and temperatures between 300 and 600 K are indicated by the *black* hatched ellipse. Above the *dashed line* bulk RuO_2 is thermodynamically unstable against CO-induced decomposition (see text). Adapted from [60]

As long as the (T, p_i)-conditions of two experiments correspond to the same μ_i, thermodynamically the same results would be expected. In order to represent the chemical potentials of (near-)ambient catalysis in UHV Surface Science one would correspondingly have to resort to much lower temperatures, cf. the different pressure scales in Figs. 7.2 and 7.3. Note, however, that this idea of thermodynamic scaling by maintaining the same chemical potentials is not necessarily the same as simply maintaining a constant reactant partial pressure ratio and varying the total pressures or temperature. Such a procedure does not keep the chemical potentials constant, and in case of dissociatively adsorbing reactants generally not even the chemical potential ratios. Without knowledge of the surface phase diagram, the concomitantly explored chemical potential range may easily cross phase boundaries, and then lead to incomparable results even on thermodynamic grounds alone.

Obviously, also kinetic limitations will contribute to deviations from thermodynamic scaling and further jeopardize a reliable bridging of the pressure gap by simple thermodynamic recipes for the gas-phase concentrations [10]. Such kinetic effects are thereby not necessarily more prominent at low temperatures. At higher temperatures one may generally expect higher turnover frequencies. The surface-reaction processes might thus increasingly occur at higher rates than the adsorption and desorption processes that maintain the equilibrium with the surrounding gas phase that is assumed in (constrained) ab initio thermodynamics. As further discussed below, the resulting depletion of particular surface species may then well lead to significant deviations from the predicted surface structure and composition [60]. Already for the pure formation of the thin surface oxide overlayer on Pd(100) at increasing O pressures, in situ surface X-ray diffraction (SXRD) experiments indicated severe kinetic limitations that suppressed formation of the overlayer at near-ambient pressures and elevated temperatures on the time scale of hours [42]. One may well imagine such limitations to intensify in the presence of a reducing co-reactant, or when formation of thick bulk-like films is concerned. This should be kept in mind when assessing the results of Fig. 7.3. For the more reactive Ru metal, the stability region of its bulk RuO_2 oxide is much larger than for Pd and PdO [47]. In terms of the surface phase diagram, technologically-relevant reaction conditions fall therefore well into the stability region of this bulk oxide, cf. Fig. 7.3. Instead of a potential (dynamic) formation of a nanometer-thin surface oxide overlayer as on Pd(100), this would rather suggest thick bulk-like oxide films to occur on Ru, with the catalytic phase coexistence then restricted to the adsorbate overlayer on these films. However, kinetic growth limitations, e.g. due to slow diffusion of either O or Ru atoms through the formed film [63, 64] might significantly change this picture. Indeed, while the formation of crystalline, bulk-like $RuO_2(110)$ during (near-)ambient CO oxidation catalysis has indeed been observed experimentally at Ru(0001), even after long operation times the reported film thicknesses never exceeded about 20 Å [13, 65, 66].

This restates to really view the results of (constrained) ab initio thermodynamics only as very rough first insights. However, even on this level these insights can be very valuable and in the discussed context of oxide formation in (near-)ambient oxidation catalysis these insights do support the dynamical catalysis picture in terms of substantial surface morphological transformations in the reactive environment that has emerged from corresponding in situ experiments. In fact, as there is no reason why a possible formation of (surface) oxides should simultaneously occur on different facets of the same metal, such transformations can also contribute to substantial changes in the shape and morphology of (supported) nanoparticles. (Constrained) ab initio thermodynamics can also contribute to this context by calculating surface free energies of different facets and combining them within Wulff (Kaischew) constructions [67, 68]. Significant particle shape changes have this way indeed been predicted as a function of the surrounding gas-phase environment [69–72]. The possibility to quickly compare surface configurations that vary as widely as metal, oxidic overlayer, and bulk-like oxides is thereby an asset that—behold of the highly approximate nature of this theory—cannot be overstated and that serves ideally to elucidate the dynamics of working catalysts.

7.3 First-Principles Microkinetics

7.3.1 Methodology

In order to properly capture the kinetic effects that are suspected to modify the approximate picture obtained within ab initio thermodynamics, the simulations need to explicitly account for a time dependence. The involved time integration is thereby extensive and may exceed time scales of the order of seconds. The reason for this is the so-called rare-event dynamics underlying surface catalytic processes.

While a catalyst generally reduces the barriers of these processes, they are still typically of the order of ~1 eV. Since this is much larger than k_BT, the time scales of these relevant elementary processes are largely decoupled from the regular thermal (vibrational) motion. A vanilla-flavor molecular dynamics simulation integrating the Newtonian equations of motion for the nuclei would be able to capture these vibrations. Yet, it is largely intractable to integrate up over time scales that would allow for a statistically relevant averaging of the rare catalytic processes.

In microkinetic theories this separation of time scales is instead exploited by abandoning the continuous dynamical description in favor of a discrete state-to-state time evolution, in which the individual elementary processes drive the system in discrete jumps from one system state to the next [73, 74]. The central equation to solve is then a Markovian master equation

$$\frac{dP_\alpha(t)}{dt} = \sum_\beta \left[W_{\alpha\beta} P_\beta(t) - W_{\beta\alpha} P_\alpha(t) \right], \tag{7.4}$$

where α and β are states of the system with corresponding probabilities $P_\alpha(t)$ and $P_\beta(t)$. $W_{\alpha\beta}$ and $W_{\beta\alpha}$ are the transition probabilities per unit time, specifying the rate with which the system changes due to the elementary processes (adsorption, desorption, reaction, and diffusion), respectively from state β to α and vice versa. These master equations, one for each system state α, are thus simple balancing equations: The probability to find the system $P_\alpha(t)$ in state α at any time t changes because transitions from any other state β can occur into state α ($W_{\alpha\beta}P_\beta(t)$) or they can occur out of state α into any other state β ($-W_{\beta\alpha}P_\alpha(t)$). Importantly, one has thereby applied a Markov approximation, because none of these transitions depend on the history through which states the system has gone before. Rather than involving probabilities that depend on any past time $t' < t$, (7.4) thus only shows probabilities at the same instant in time t: Transitions involving a hopping out of state α at time t depend only on the probability that the system actually is in state α at time t ($P_\alpha(t)$). Transitions involving a hopping from any other state β into state α at time t depend only on the probability that the system is in state β at this time ($P_\beta(t)$). The rationale behind this approximation is that one assumes that during the long vibrational motion before a rare event eventually brings the system out of the current state into the next one, the system completely forgets how it actually got into this state in the first place. Limitations in the dissipation of the reaction energy released during individual elementary processes might potentially lead to violations of this Markov approximation [15], but for the time being this approximation is unanimously assumed in prevalent formulations of chemical kinetics.

For a small number of system states, a Markovian master equation like (7.4) can be solved analytically. Unfortunately, in surface catalysis we are not facing such a small number. On the contrary. Assume that our catalyst surface exhibits a total of N active sites. A unique system state would then be defined by the detailed population of every single one of these sites, and any elementary process that changes the population of one or more of these sites corresponds to one entry in the transition matrix $W_{\alpha\beta}$ [75]. Since the examples I use are for CO oxidation catalysis, let's stick to this reaction to see what this means in terms of numbers. In this reaction, any active site can either be empty, or occupied by the reaction intermediates O or CO (if we assume that CO_2 formation leads to immediate desorption of the product). This yields three population possibilities for every site and if we for example assume that there are $N = 100$ active sites, then the total number of system states, also known as detailed population configurations of the sites, is $3^{100} \approx 10^{47}$. Obviously, this is not a small number and for any more complex reaction network with a correspondingly increased number of different reaction intermediates it will even be higher. Yet, we still have to rationalize why $N = 100$ should be a good representation for an extended catalyst surface. This comes about as in order to appropriately capture the ensemble effects at such a surface, the explicitly considered group of active sites (that is suitably continued through periodic boundary conditions) must be large enough to exceed the correlation length between sites. This is the length over which the statistics of the processes that are ongoing at one site still influences the statistics of the processes that occur at another. From present experience on the type of systems discussed in this review an area spanned by $(10 \times 10) = 100$ sites is a good (in fact lower) estimate for this [22, 75].

For a surface catalytic system we thus have to generally expect a transition matrix with a dimension of the order of $\sim(3^{100} \times 3^{100})$ or larger, i.e. with at least $\sim(3^{100})^2 \approx 10^{95}$ matrix elements. Fortunately, most of these matrix entries are zero [75]. This has to do with the fact that chemical elementary processes typically affect only the population of a small number of sites. A unimolecular adsorption or desorption event of a molecule like CO changes the occupation of one particular site. A diffusion process of such a molecule changes the occupation at two active sites, one being emptied and an empty one being filled. Any transition connecting states that differ in their populations by more than a few individual sites has therefore a $W_{\alpha\beta} = 0$. An additional important feature that simplifies the transition matrix immensely is translational symmetry at a crystalline extended surface. In such a situation our ensemble of $N = 100$ active sites may only comprise a much smaller number of inequivalent site types. At a simple low-index metal surface maybe something on the order of two or three, say hollow or bridge terrace sites or high-symmetry sites at an upper or lower step edge. In the crystalline symmetry the elementary processes occurring at any site type are equivalent, which means that their corresponding transition matrix elements $W_{\alpha\beta}$ are the same. While the total number of non-zero matrix elements even in the largely sparse transition matrix is thus generally still too large to even be stored, the total number of inequivalent matrix entries $W_{\alpha\beta}$ is then typically rather small and determined by the total number of inequivalent elementary reactions in the reaction network [75–77]. For a simple CO oxidation model comprising only one active site type this total number can in fact be as low as seven: Dissociative adsorption of O_2, associative desorption of two adsorbed O, CO adsorption, CO desorption, O diffusion, CO diffusion, and CO + O reaction. It is only this immense simplification due to a prevailing and despite the ongoing catalytic reactions static crystalline symmetry that makes any kind of microkinetic model computationally tractable today. I come back to this point later, but already here we should realize that this obviously clashes with our working hypothesis of a dynamically evolving, possibly amorphous or highly heterogeneous catalyst surface that we intend to scrutinize with such simulations. This is precisely the dilemma. We are largely constrained to conduct microkinetic simulations within static models focusing on impinging and reacting gas-phase species on a rigid solid surface. In a self-fulfilling prophecy this then contributes to the present widespread acceptance of such a picture of catalysis.

Even though the transition matrix is thus sparse and contains only few non-equivalent non-zero matrix elements, this does not change the fact that its dimension is of the order of $\sim(3^{100} \times 3^{100})$ already for the discussed CO oxidation reaction. While in the notation of (7.4) the master equation has a deceptively simple form, it is hence so high-dimensional that it generally escapes any direct solution. Kinetic Monte Carlo (kMC) simulations overcome this problem by generating an ensemble of state-to-state trajectories with the property that an average over the entire ensemble of trajectories yields the probability densities $P_\alpha(t)$ of (7.4) [21, 22]. In this way, only those matrix elements $W_{\alpha\beta}$ of transitions between states α and β are required that are actually executed along the generated trajectories. Despite the averaging over the trajectory ensemble, a central feature of the finally obtained explicit numerical solution is thereby that it contains the information of the detailed spatial distribution of the reaction intermediates over the considered active sites, along with the equally resolved occurrence of the individual elementary processes. The still prevailing alternative to achieve a solution of (7.4) discards this detailed information and instead considers only the occupation probabilities at different site types, i.e. the averaged coverage θ of all equivalent sites of a given type [5–7]. This represents a significant simplification of the problem, as the master equation then decays into a small number of differential rate equations describing the time evolution of these coverages at the different site types [78, 79]. These are exactly the type of rate equations that are often phenomenologically formulated. Typically the resulting network of differential equations is extremely stiff and requires special solution techniques. Nevertheless, even then the computational solution is so undemanding that it can mostly be achieved on time scales of the order of seconds on simple desktop computers.

There is also an additional simplification with respect to the input that such a mean-field (MF) rate equation model requires. It only needs to know what kind of active site types are considered and which elementary processes can take place at each one of them. In contrast, as it resolves the spatial distributions at the surface, a kMC model additionally needs to know how these active site types are geometrically arranged with respect to each other. As already mentioned such simulations are presently only tractable under a prevailing translational symmetry. Typically, kMC simulations in the field are therefore performed for a given lattice model that reflects the crystalline structure of the studied single crystal surface or nanoparticle facet. From this perspective, and recalling our objective to investigate a possible dynamical picture of catalysis, this sounds like a disadvantage or limitation in comparison to the MF rate equation approach. To some extent this is true. On the other hand, one has to realize that an MF model does not even know whether there is a crystalline order at the surface or not. It does not even know that step sites are per definition linearly coordinated next to each other and are thus differently accessible to surface reaction intermediates than active sites at a two-dimensional terrace. The only thing an MF model knows and can correspondingly account for is that there are the different active site types that it considers. Obviously, MF rate equation theory is thus a gross approximation in comparison to kMC and we can only expect it to yield a faithful description of the surface kinetics if this approximation is justified. The latter is the case, when there is a perfect mixing of the reaction intermediates over the active sites of the surface. Then, indeed, the details of the spatial distribution do not matter. Fast diffusion processes can ensure such a mixing. In turn, diffusion limitations, as we can often expect them for example at oxide surfaces, are one of the two classic situations known to cause a break-down of the MF approximation, with rate equation theory correspondingly providing inaccurate solutions [79, 80]. The other situation arises in the case of strong lateral interactions between reaction intermediates, as the implied preferences of certain reaction intermediates to either seek or avoid each other naturally oppose the diffusional tendency to randomly mix the adlayer [78]. As it is not a priori obvious if the MF assumptions are fulfilled for a given system, MF rate equation theory should not be applied uncritically. Clearly, if they are fulfilled, MF theory is the much more efficient approach that should be pursued. If they are not fulfilled, wrong results and concomitant "ideas" might result.

KMC and MF rate equations are presently the two predominant microkinetic theories. As rate equations are the far more traditional and widespread approach, people often exclusively associate them with the label microkinetic modeling. This is sloppy as both theories formally provide solutions to the same microkinetic master equation. With the rapidly advancing use of kMC simulations in the field of surface catalysis one should thus rather refer to microkinetic modeling as a joint label for both approaches. The formal similarity of the two approaches is also reflected in the equivalent input they require. As already discussed these are the inequivalent active sites (in kMC additionally a lattice model fixing their geometrical arrangement) and the list of elementary processes that can occur at these sites. It is worthwhile to emphasize that this is an input, not an outcome of the simulations. Neither approach has any built-in warning feature if a relevant process or site type has been overlooked, or even more desirable the capability to automatically generate complete lists of such processes and sites. If a relevant process or site is not included in the microkinetic model, the results are nothing, but simply just wrong.

7.3.2 The First-Principles Input

Apart from these lists the remaining input that is additionally needed are the inequivalent, non-zero transition matrix elements. With units of time^{-1}, these matrix elements correspond to the rate constants of the various elementary processes, i.e. $W_{\alpha\beta} = k_a$ if the transition from state α to state β results from elementary process a with rate constant k_a [22, 75]. In first-principles (1p) microkinetic approaches these rate constants are determined by electronic structure theory calculations, and it is through these rate constants that such kind of modeling then obtains its (hopefully) predictive character. To derive the rate constants predominantly from computationally less demanding static, again typically DFT, calculations, the currently most commonly employed approach in the area of surface catalysis is transition-state theory (TST) [19, 81, 82]. Without having seen much systematic scrutiny, this approach seems to meet sufficient accuracy, which is thus quite different to the situation in other fields e.g. when liquids are involved. TST yields rate constants of a general form

$$k_a = A(T, p_i)\exp\left(\frac{-\Delta E_a}{k_B T}\right), \tag{7.5}$$

where the prefactor $A(T, p_i)$ accounts for entropic changes between the initial and transition state (TS) of the process, and ΔE_a is the corresponding energy barrier. As the prefactor enters this equation only linearly, various, in parts drastic, approximations for it characterize the present state-of-the-art in the field [19, 83]. In particular for adsorption or desorption processes or Eley-Rideal reaction steps that may involve large entropy changes this will have to be improved in future work [60, 84]. Apart from their direct quantitative impact on the rate constant and subsequently the microkinetic simulation result, such approximations have generally also to be seen in the light of microscopic reversibility. In order to be thermodynamically consistent, rate constants of forward and (time-reversed) backward processes like adsorption and desorption have to fulfill a detailed balance condition. If different approximations are made for the two processes, this condition can be broken. Kinetic models that correspondingly do not yield the proper thermodynamic limits should be met with great skepticism, but are unfortunately frequently found in the literature.

This leaves as most crucial DFT input the energy barriers ΔE_a for every inequivalent elementary process a. Already for decently sized reaction networks and considering only a few inequivalent active site types, the explicit calculation of these barriers quickly becomes the predominant computational bottleneck of 1p microkinetic studies [19]. This in particular, as the ΔE_a generally depend on the local environment, i.e. lateral interactions with nearby co-adsorbates modify the energy barriers. In order to capture such effects, multiple DFT calculations of the same process need to be performed for different local adsorbate configurations. In 1p-kMC simulations these are then cast into some (short-range truncated) lattice-gas Hamiltonian expansion [85–88], while in 1p-MF rate equation theory this dependence is considered through an effective coverage-dependence $\Delta E_a = \Delta E_a(\theta)$ [7, 89, 90]. In their prevalent formulation 1p microkinetic studies thus carry an enormous overhead. Extensive DFT calculations are required to determine all process barriers and their environment dependencies. This information is then stored in look-up tables, which serve as basis for the subsequent and computationally typically far less demanding actual 1p-kMC or 1p-MF rate equation simulations. An obvious disadvantage of such a static divide-and-conquer type procedure is that potentially extensive DFT calculations are performed for reaction intermediates or coverage regimes that in the actual microkinetic simulations for the targeted reaction conditions are never met.

A pragmatic solution to this is to start with quite simple formulations for the reaction network and lateral dependencies, possibly using lower-level theories for an only approximate account of the lateral interactions. In a second step one iteratively refines the model depending on the simulation results one obtains. Due to the non-linearities of the reaction network, such an approach is not fool proof though, i.e. the initial model can be so coarse that it leads into a completely wrong direction. A highly appealing alternative especially for the trajectory-based 1p-kMC simulations would therefore be to only compute the really required reaction barriers on-the-fly, i.e. in the course of the on-going 1p-kMC simulation. Such approaches come with names like adaptive kMC, on-the-fly kMC, self-learning kMC, or kinetic activation-relaxation technique [91–94]. They would indeed also be most appealing from the perspective of a dynamical catalysis picture, as such approaches would not necessarily be restricted to a fixed lattice model. The essential idea of these kind of on-the-fly kMC formulations is to compute all energetically low-lying (and therefore dynamically relevant) barriers out of a given system state α. In accordance with the kMC algorithm, one of the corresponding elementary processes is executed and brings the simulation into a new system state β. This process is then iterated, i.e. barrier calculations are performed sequentially for every new state visited. Huge savings in computational time can thereby be achieved when appropriately storing the already computed barriers and introducing some form of state recognition. If the algorithm thus realizes that the new state β corresponds to an already visited earlier state α, barriers are not recomputed, but drawn from the existing look-up tables. Despite these savings, the computational effort of an at least semi-reliable exploration of all low-lying barriers at individual kMC steps is generally still orders of magnitude higher than those of the traditional divide-and-conquer look-up formulation. Applications of on-the-fly kMC in surface catalysis are therefore presently either restricted to very specialized systems with only reduced configurational complexity or they employ force fields rather than DFT calculations for the process barriers.

The computationally expensive part of an actual barrier calculation is in either case the location of the TS through advanced transition state search algorithms [95, 96]. In on-the-fly 1p-kMC, where the final states are not known, this would be one-ended techniques like the dimer method [97, 98]. In the prevalent divide-and-conquer 1p microkinetic approaches, where initial and final state of an elementary process are known, most accurate results are instead obtained by state-of-the-art two-ended techniques like the (climbing image) nudged elastic band (NEB) method or string approaches [99–101]. Regardless of dimer, NEB, or string, one TS search will involve numerous individual DFT calculations. For the system sizes typical for surface catalytic problems these DFT calculations may furthermore exhibit severe convergence issues, or the actual TS search algorithm has problems converging to the (right) TS. The barrier determinations are therefore the by far most critical and (human and CPU) time-consuming step in a 1p microkinetic study. Obviously, it is thus also this step that has the highest leverage for speed-ups through more approximate approaches. This starts already with the use of less rigorous TS search algorithms like drag methods or a mere calculation of energy profiles along assumed reaction paths. However, most prominently and with highest efficiency gains, this has been exploited by approximate relations between the activation energies and the thermochemistry of the reaction [102–108]. One prominent example are the well-known Brønsted-Evans-Polanyi (BEP) relationships [5, 7, 106–108], which yield linear relations of the kind $\Delta E_a \approx c_1 (E_f - E_i) + c_2$, where c_1, c_2 are constants and $(E_f - E_i)$ is the energy difference of the initial and final state of the reaction. Since the latter thermochemical energy difference only involves geometry optimizations of (meta)stable configurations, knowledge of such a relation yields substantial reductions in computational cost as compared to an explicit TS search. An even further reduction in cost and the number of independent parameters has been achieved by realizing that the binding energetics of many reaction intermediates can be related to the binding energetics of a few base elements out of which these reaction intermediates are typically composed, namely H, C, N, O, and S [108–110]. While the initial task was thus to explicitly compute a considerable number of energy barriers for each elementary process of the considered reaction network, exploitation of the latter scaling relations and BEP relations may reduce this to the calculation of the binding energies of a few base elements. This can imply such an enormous reduction in the computational cost that it allows to access quite complex reaction networks and in particular engage in computational screening studies [7, 108, 111–119]. This route has hitherto been exclusively pursued within 1p-MF rate equation approaches. As the goal of kMC-based 1p microkinetic modeling is typically more a comprehensive and most accurate understanding of a particular system, use of such more approximate scaling and BEP energetics may have seemed less obvious. However, there is no conceptual obstacle against doing so in the future.

7.3.3 Surface Morphological Transitions in Near-Ambient Catalysis

Just as with (constrained) ab initio thermodynamics, a central outcome of 1p microkinetic modeling is the surface structure and composition as a direct function of the surrounding gas phase. As the theory is explicitly time dependent, this can be for steady-state reaction conditions, but equally for non-stationary situations as for example in temperature-programmed-reaction (TPR) experiments. 1p-MF rate equation theory provides this information in the form of average coverages at the considered active sites. 1p-kMC simulations additionally provide the detailed spatial distributions and fluctuations at the surface. Such insight is invaluable to properly capture and analyze microstructural effects, for instance at oxide surfaces or defects like vacancies or steps. Of course, the 1p-kMC distributions can also be averaged to obtain (proper) average coverages without having to resort to the MF approximation.

In the resulting surface populations the kinetic effects due to the on-going reaction events (that were neglected in constrained ab initio thermodynamics) are now explicitly considered. Also, "phase" transitions are better described as "configurational entropy" is accounted for. In 1p-MF rate equation theory without any coverage dependencies this is at a level equivalent to Langmuir models [120], in 1p-kMC this is the accurate numerical evaluation on the ensemble of active sites considered. Quite deliberately, I have put the words "phase" and "configurational entropy" in quotes here, as these are inherently thermodynamically defined terms, while the consideration of an open catalytic system with on-going reaction events in 1p-kMC and 1p-MF obviously brings us outside the realm of thermodynamics. To reflect this, pioneering kMC work on surface catalytic problems [121] has created the, sometimes critically mocked word "kinetic phase diagrams" (then containing "kinetic phase transitions" etc.) to denote the equivalent compositional output as compiled in the surface phase diagrams of (constrained) ab initio thermodynamics as e.g. shown in Figs. 7.1, 7.2 and 7.3. In the following I will stay within this type of nomenclature in exactly the spirit as put forward by Ziff, Gulari, and Barshad [121].

Figure 7.4 shows such a kinetic phase diagram for the CO oxidation problem at $RuO_2(110)$ that I discussed at the (constrained) ab initio thermodynamics level above. Directly compared are results obtained by 1p-kMC simulations and 1p-MF rate equation theory [60, 78]. Both microkinetic simulations have been based on exactly the same DFT input and the same considered reaction network, such that the differences discernible in Fig. 7.4 arise exclusively from the mean-field approximation in the MF approach. Even though the overall topology of the phase diagram is largely robust against this approximation, the positions of the catalytically most relevant kinetic phase boundaries are somewhat shifted. A detailed analysis shows that this goes hand-in-hand with significant shortcomings of MF theory to appropriately describe the catalytic activity and underlying reaction mechanisms [78]. More important for the present context are, however, the much more significant deviations in the predicted surface structure and composition when comparing both 1p microkinetic theories with the approximate thermodynamic insight in Fig. 7.3. What prevails is the insight that technologically relevant reaction conditions with pressures of the order of 1 atm and near-stoichiometric reactant ratios fall in the vicinity of a phase transition, and in particular the one in which adsorbed O and CO compete for the so-called coordinately unsaturated (cus) sites offered by this surface. This finding and the importance of the cus sites for the catalytic activity of $RuO_2(110)$ are fully consistent with all presently available experimental data [11, 13, 14]. Substantial differences between (constrained) thermodynamic and microkinetic theory are, however, obtained for the population of the other (br) idge active site type offered by the $RuO_2(110)$ surface. While (constrained) ab initio thermodynamics predicts a predominant population with O^{br} even for largely CO-rich gas-phase conditions [37, 59], both microkinetic theories agree on an essentially complete replacement by CO^{br} species in this regime. This is a classic illustration of the surface catalytic reactions consuming a reaction intermediate, here O^{br}, faster than it can be replenished by adsorption from the gas phase. Since ab initio thermodynamics is blind to such kinetic effects, it only assesses the very strong binding of O to these bridge sites and thereby largely overestimates the presence of this species at the surface.

This showcase example thus nicely underscores the approximate nature of (constrained) ab initio thermodynamics results and the added value of explicit 1p microkinetic theories. Of course, not everything is perfect in the latter theories either. Even in the highly CO-rich gas-phase conditions in the upper left parts of the panels in Fig. 7.4 both 1p microkinetic theories predict at maximum a fully CO-poisoned oxide surface, whereas the thermodynamic estimates in Fig. 7.3 immediately reveal the proper complete reduction of the oxide. This difference arises as the predictive power of the 1p microkinetic approaches extends, of course, only to the active sites and concomitant set of elementary processes considered in the model. In the studies behind Fig. 7.4 this framework corresponded to the active sites of a reduced, but otherwise intact $RuO_2(110)$ surface. The structural complexity that would arise when considering a full oxide reduction path would presently imply a completely intractable 1p input (vide infra), let alone that at best only a conceptual perception of the individual mechanistic steps involved in such a path is available to date [122, 123]. For the targeted CO oxidation activity of $RuO_2(110)$ this limitation with respect to oxide reduction is thereby not actually the real problem. Relevant, near-stoichiometric gas-phase conditions are located sufficiently well inside the stability regime of the bulk oxide, cf. Figs. 7.3 and 7.4. However, a long-term deactivation of this $RuO_2(110)$ facet has been experimentally reported even for oxidizing feeds, which was assigned to a microfaceting into an inactive c (2×2)-$RuO_2(100)$ structure [124]. Again, such a reaction-induced complex

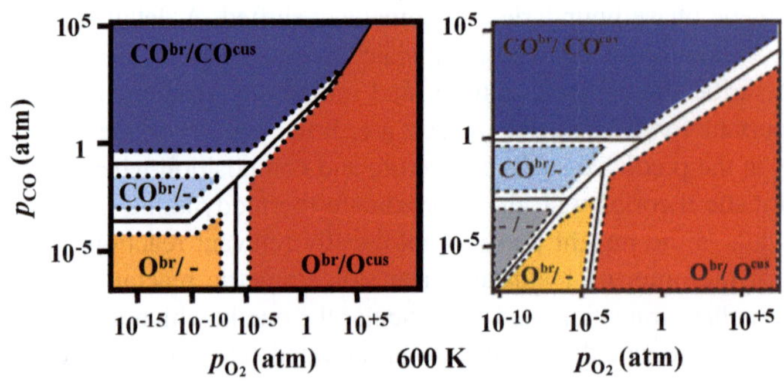

Fig. 7.4 Kinetic surface phase diagrams for the $RuO_2(110)$ surface in an environment consisting of O_2 and CO at 600 K. Compared are results from 1p-kMC simulations (*left panel*) with results from 1p-MF rate equation theory (*right panel*). The labels of the different stable phases reflect a predominant population (O, CO, or empty "–") of the two prominent adsorption sites offered by this surface, br(idge) and coordinatively unsaturated (cus) site. Coexistence regions at the phase boundaries are marked in *white*. Labels and shown ranges of partial pressures (gas-phase chemical potentials) are identical to those in Fig. 7.3. From [60, 78]

surface morphological transition—which is a prototypical example for exactly the dynamical view of an evolving catalyst we would like to scrutinize—is presently largely outside the reach of predictive-quality microkinetic modeling capabilities.

Fortunately, the situation is a bit more accessible for the Pd(100) example discussed before. Due to the reduced stability of bulk PdO, here "only" the (possibly continuous) formation and reduction of a thin surface oxide film while on stream is to be assessed. A first step in this direction has been taken by simply performing 1p-kMC simulations once on the pristine metal, i.e. for a lattice model and set of elementary reactions pertinent to Pd(100), and once on the perfectly intact surface oxide, i.e. for a lattice model and set of elementary reactions pertinent to the $(\sqrt{5} \times \sqrt{5})R27°$ surface oxide [87]. Evaluating the average surface composition for a wide range of gas-phase conditions one can assess the boundaries within which one would still trust either of the two models. Detailed experimental work indicates the onset of surface oxide formation once a critical O coverage around and above 0.5 ML on Pd(100) is exceeded [46]. This suggests the 1p-kMC Pd(100) model as a faithful representation for gas-phase conditions where the O coverage stays well below this value. Equivalently, one would expect the onset of surface oxide decomposition whenever a critical coverage of surface oxygen vacancies, say 10%, is exceeded [44]. For gas-phase conditions where this coverage is much lower, the intact 1p-kMC surface oxide model should be a good representation. Intriguingly, the results of the corresponding 1p-kMC simulations shown in Fig. 7.5 identify a finite range of (T, p_{CO}, p_{O2})-conditions where both stability criteria are fulfilled [87]. In this range the Pd(100) 1p-kMC model predicts an O coverage below 0.25 ML, while simultaneously the $(\sqrt{5} \times \sqrt{5})R27°$ surface oxide 1p-kMC model predicts a surface oxygen vacancy concentration well below 10%. The corresponding bistability region is thereby quite robust against uncertainties in the DFT energetics or the treatment of lateral interactions. Moreover, its location in (T, p_{O2})-space in fact comprises precisely the near-ambient reaction conditions for

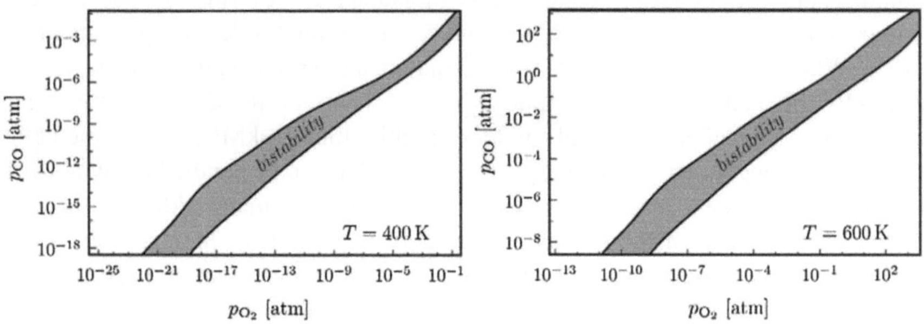

Fig. 7.5 Bistability region in CO oxidation catalysis, i.e. gas-phase conditions where 1p-kMC models simultaneously predict the stability of pristine Pd(100) and the $(\sqrt{5} \times \sqrt{5})R27°$ surface oxide. At 600 K this bistability region comprises technologically relevant (near-)ambient, stoichiometric gas-phase conditions. At 400 K this region is shifted to more O-rich conditions as employed in the Reactor STM experiments by Hendriksen et al. [56, 61, 62]. From [87]

which Reactor STM studies had reported an oscillatory formation and decomposition of an oxidic film at the surface of the working catalyst [56, 61, 62].

In particular at elevated temperatures, 600 K in Fig. 7.5, the bistability region centers on technologically most relevant near-stoichiometric partial-pressure ratios. These findings thus fully support a dynamic view of catalysis at least in the sense that a surface morphological transition, here the formation of a thin surface oxide layer, may indeed occur in the reactive environment. The simulations, performed separately on the two intact surface states, can, however, not address whether the very dynamics of the transition itself is a key factor. In other words, whether it is only the continuous formation and decomposition of the oxidic film in the experimentally reported oscillations that creates the real active sites, e.g. in form of transient structures or at domain boundaries on the evolving surface. For this the 1p-kMC simulations would have to be able to represent both surface states and transitions between them. For this very system a step in this direction has in fact recently been taken through a novel multi-lattice kMC approach, which exploits the lattice commensurability of the $(\sqrt{5} \times \sqrt{5})R27°$ surface oxide with the Pd(100) surface [125]. The latter allows to establish a superlattice model that simultaneously comprises both metal and surface oxide sites, with the multi-lattice kMC algorithm keeping track of which surface areas are either in the oxide or the pristine metal state by appropriately activating or deactivating elementary processes at the corresponding sites. At present this approach has only been applied to the reduction of the surface oxide in a CO atmosphere [125]. Intriguingly, CO oxidation reaction steps across metal-oxide domain boundaries turned indeed out to be essential to reproduce the experimentally reported temperature dependence of the reduction rate [126].

Whether the same or other processes related to the dynamics of an evolving surface are also crucial for steady-state CO oxidation catalysis remains yet to be seen. The price to pay for such insight through multi-lattice kMC simulations is to establish a detailed atomistic pathway for the transition between the treated system states, here the pristine metal and the surface oxide. The exploitation of the lattice commensurability renders this endeavor tractable. It nevertheless constitutes a computationally most expensive step involving a multitude of 1p calculations [125]. While this obviously restricts the dynamical phenomena in catalysis that can presently be tackled, an important aspect to keep in mind is the following. Regardless of whether traditional single- or multi-lattice 1p-kMC, already the lattice models and concomitant elementary process lists that can be handled today allow to treat quite complex reaction networks that comprise many different reaction mechanisms. Which of these reaction mechanisms dominates the catalysis is then an output of the simulations, not an input. This is a crucial asset that distinguishes such 1p microkinetic simulations from ubiquitous kinetic studies where a certain reaction mechanism is simply assumed, often based on rather little or only indirect evidence.

7.3.4 Catalytic Activity from First Principles

Another important asset of 1p microkinetic simulations is, of course, that they do not only provide information about the surface structure and composition, but also determine the catalytic activity and, if applicable, the selectivity. Just as much as for the surface (kinetic) phase diagrams this information can be computed in steady state for a range of gas-phase conditions and then be compiled in corresponding, so-called TOF maps. Alternatively, if transient situations are addressed, it can for example be computed for various initial system states. Figure 7.6 shows examples for such data drawing on the previously-discussed example of CO oxidation at $RuO_2(110)$ [127, 128]. In both shown examples the absolute pressures addressed are in the UHV regime, which makes it possible to directly compare to corresponding data from Surface Science experiments (vide infra). In both cases excellent agreement is reached, which in particular for the transient TPR data is only obtained through the appropriate consideration of the spatial distributions at the catalyst surface. As shown in Fig. 7.6 qualitatively different variations with initially prepared O^{cus} coverage would be expected for two competing reaction mechanisms, $O^{br} + CO^{cus}$ (~linear variation) and $O^{cus} + CO^{cus}$ (~parabolic variation). The latter mechanism is known to be the more reactive one due to the much weaker binding of the O^{cus} species. The at first glance enigmatic strong suppression of this mechanism seen in Fig. 7.6 is instead a direct result of diffusion limitations in the trench-like arrangement of the cus sites under the specific experimental TPR conditions. Such an effect can only be captured by 1p-kMC simulations, which only then are able to reconcile the known higher reactivity of the $O^{cus} + CO^{cus}$ mechanism with the linear profile measured in the TPR experiments [128]. Both for this

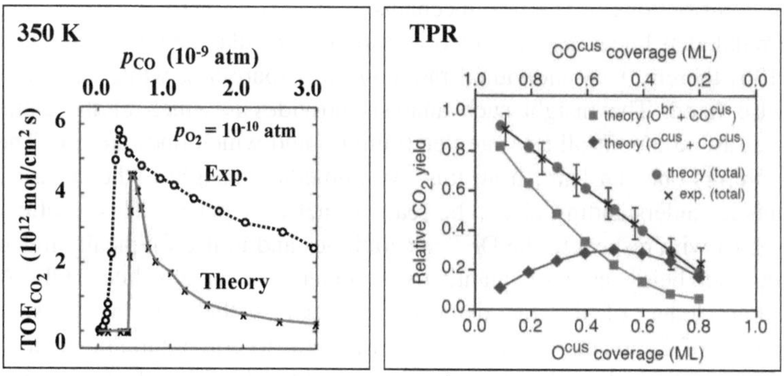

Fig. 7.6 Comparison of measured and 1p-kMC computed steady-state and transient catalytic activity for CO oxidation at $RuO_2(110)$. (*Left*) Steady-state turnover frequencies (TOFs) at 350 K. (*Right*) Total CO_2 yield obtained in TPR spectroscopy for surfaces initially prepared with varying O^{cus} and CO^{cus} coverage. The CO_2 yield is given relative to the one obtained for the surface with zero O^{cus} coverage. Shown are the total simulated CO_2 yield and the contributions from the two dominant reaction mechanisms under these conditions, $O^{cus} + CO^{cus}$ and $O^{br} + CO^{cus}$. From [127, 128]

example and in general, the capability to explicitly resolve the contributions of individual reaction mechanisms to the overall (and observable) catalytic activity is thus a most important aspect for the mechanistic understanding. Obtaining wrong relative contributions correspondingly bears the risk of deducing wrong conclusions (and "ideas"). Similar to the TPR case, a wrong ordering of the contribution from different reaction mechanisms in 1p-MF rate equation theory has also been reported for steady-state reaction conditions [78], which thus adds to the list of shortfalls of this theory if the MF approximation is unjustified.

Despite reports of a number of similarly successful 1p microkinetic studies [113], one has to recognize that reaching a quantitative agreement in absolute TOFs cannot generally be expected. This holds already because of the typically large uncertainties in experimental absolute TOFs. The uncertainties on the theoretical side are not any smaller, primarily due to the aforementioned uncertainties in the approximate DFT energetics. At the temperatures of interest in catalysis, the quoted ~0.3 eV (~30 kJ/mol) uncertainty in DFT barrier values translates into 1p rate constants that can be wrong by several orders of magnitude. For more approximate BEP or scaling-derived barriers, this will be even worse. At first glance such a large uncertainty seems to invalidate any attempt to compute meaningful TOFs, or it lends support to the pragmatic approach to empirically "correct" 1p microkinetic simulations such that they match certain experimental findings [129]. A more constructive approach that does not sacrifice the invaluable independence of a first-principles theory is instead to systematically analyze which errors in the 1p energetic data base can really contribute to what degree to errors in the predicted activity (or other properties of interest). A central concept in this respect are so-called sensitivity analyses, which loosely speaking are nothing but a systematic variation of the input energetic parameters to assess the influence this has on the outcomes of the microkinetic model (surface composition, activity, selectivity, relative contributions of reaction mechanism, etc.) [130–136].

Formulated as linear response theories, approaches like the degree of rate control [131, 135] thereby vary individual rate constants (barriers), while keeping everything else fixed. The insight such analysis provides is which of the elementary processes are rate-controlling (rate-determining) and which ones are not. There are several things one can learn from this. An immediate insight is the corresponding mechanistic understanding about the reaction network per se. This is often much more robust with respect to the DFT uncertainties and in itself typically much more relevant than being able to quantitatively determine an absolute TOF. Among others knowledge of the rate-determining steps is the gateway to simplified descriptions of the reaction network and therewith to computational screening, as much as it identifies those kinetic bottlenecks that need to be addressed in a rational design of improved catalysts. With respect to the DFT uncertainty, rate constants of not- rate-determining processes can typically be varied by several orders of magnitude without having any effect on the simulation result. We correspondingly learn that DFT errors in such rate constants are irrelevant. On the contrary, any error contained in the description of rate-determining steps will directly propagate through and these are then the errors one should worry about.

In some cases knowing which energetic input quantities are the crucial ones already allows to rationalize agreement or disagreement with experimental data. For the steady-state catalytic activity shown in Fig. 7.6 a degree of rate control analysis e.g. reveals that under the probed gas-phase conditions it is primarily the dissociative adsorption of O_2 into a cus site pair that is rate determining [135]. The good agreement with experiment then comes about as this is a non-activated process. Rather than by the possibly inaccurate DFT rate constant, the TOF is in this case controlled by the limited availability of free site pairs for O_2 adsorption which is determined on the statistical-mechanical level. In the general case, sensitivity analyses identify those microscopic input quantities on which attention should be focused, say a particular binding energy or a particular reaction barrier. Benchmark against higher-level theory or experiment can then in principle provide an assessment how much the particular DFT quantity is actually in error, and through the sensitivity analysis how much this propagates through to the absolute TOF. The latter step is important as it tells, whether a deviation between simulated and experimental TOFs is really (exclusively) due to an inaccuracy in the underlying 1p energetics. As I will illustrate further below, there can be multiple other reasons for such deviations. This alone is an important argument against simply empirically "correcting" the microkinetic simulations by fitting selected 1p energetic values to match experimental activities or other meso-/macroscopic observables. Such a fudging can easily mask the true reasons for the deviation between 1p theory and experiment. Also replacing the 1p energetic quantity with a corresponding experimental microscopic benchmark quantity is a dangerous endeavor. Even if experimental quantities carry microscopic names like "adsorption energy" or "reaction barrier" they are typically the result of some approximate data analysis scheme, for which the multitude of TPR analyses represents a prominent example [137]. Rather than clean data, such numbers are thus effective quantities that contain an unspecified systematic error that is not covered by the quoted statistical error bars. Even in case of allegedly direct energetic measurements like microcalorimetry, firmly believed reference numbers do change with time and it remains an ongoing challenge to fully establish a safe experimental database for adsorption energetics [138].

A further argument against selectively replacing individual DFT energetic parameters with empirical numbers are the systematic trends often exhibited by DFT errors, with the widespread PBE functional [139] for instance suspected to show a systematic overbinding at metal surfaces [140]. Replacing individual energetic quantities breaks such trends and thereby a potential compensation of the systematic errors. Such correlations in the underlying energetic data base could also not be captured by the above-described linear-response type sensitivity analyses. In this respect, the concept behind the recently introduced Bayesian error estimation functionals (BEEF) represents an intriguing step forward [141–143]. The idea here is to generate an entire ensemble of functionals where known errors in adsorption energetics are mapped onto uncertainties of the parameters entering the electronic xc model. Rather than once, a 1p microkinetic simulation is then run multiple times, each time with different energetic data sets obtained from an appropriate sampling of this ensemble of functionals. The spread of the results obtained provides a quantitative error estimate and first applications of this BEEF concept indeed indicate that correlations in the DFT errors significantly reduce the predicted error on calculated TOFs [144].

In fact, an even larger reduction of errors was reported when comparing TOFs calculated for different metal catalysts [144]. This is important as corresponding relative activity comparisons, also of the same catalyst for different reaction conditions, are in any case much more relevant than the computation of an individual absolute TOF for one set of reaction conditions. The increased robustness of such trends could furthermore also rationalize the success of emerging computational screening studies which rest entirely on a comparison of relative activities varying over many orders of magnitude [7, 108, 111–119]. The critical aspect here is therefore likely less the 1p energetic data base, but the rather drastic assumptions on the microkinetic level that are presently made to make such studies tractable. Even through comparing an entire series from early to late transition metals, identical reaction mechanisms are for instance simply imposed (and not evaluated as in 1p-kMC simulations). As discussed at the beginning of this section, these reaction mechanisms furthermore typically only consider a few active sites as offered by a static, bulk-truncated surface. Even though the typically obtained, volcano-shaped activity variations over a transition metal series often exhibit their peaks close to metals that are known to be good catalysts for the studied reaction, it is presently not clear if this should really be seen as a validation of the imposed mechanism. As such it is an open question whether the success of the seminal screening studies has any bearing on the issue of a static versus a dynamically evolving catalyst surface. The true answer will eventually only come from future 1p microkinetic (screening) studies in which the possibility of surface morphological transitions is explicitly contained in the employed model.

7.3.5 *Mass Transfer Limitations Under Near-Ambient Conditions*

Regardless of the already discussed uncertainties in 1p calculated TOFs, there is yet another complication when comparing them to experiment that particularly applies to the in situ context, i.e. to the quest to specifically address catalytic activities at technologically relevant near-ambient conditions. For corresponding pressures the actual flow of mass and heat through the employed reactor becomes a significant factor. In fact, especially the dedicated experimental setups employed in in situ studies of model catalysts are likely to exhibit most complex flow profiles, as sophisticated spectroscopic probes and pumps often need necessarily to be placed in the direct vicinity of the catalyst surface [8, 9]. For the intrinsically targeted reaction conditions with highest turnovers of reactants into products this can give rise to heat and mass transfer limitations, i.e. significant temperature and (partial) pressure gradients inside the reactor [16–18, 145]. The local gas-phase composition (and therewith reaction conditions) directly at the catalyst surface may then deviate significantly from the nominal reaction conditions controlled at the inlet of the reactor. Existence of such mass transfer limitations generally prevents any meaningful measurements of the catalytic activity via the standard compositional analysis at a reactor outlet or orifices placed at the reactor walls. They also prevent any straightforward comparison to 1p microkinetic simulations, unless the latter are suitably integrated into a computational framework that appropriately accounts for the concentration and flow profiles in the reactor.

Such an integration into corresponding computational fluid dynamics (CFD) simulations has a longer history for MF rate equation theory [146], but could only recently be achieved for kMC based microkinetic simulations [16–18]. Placed into the context of 1p microkinetic simulations, resulting 1p-MF–CFD or 1p-kMC–CFD multiscale modeling frameworks are in their absolute infancy. For the theme of a potentially dynamically-evolving catalyst surface they nevertheless bear exciting prospects. Up to now this discussion centered only on the possibility of surface morphological transitions at the working catalyst, with in particular the CO oxidation at Pd(100) example pointing at an intrinsic heterogeneity of the surface. This does not answer the central question as to the nature of the active sites. Is one of the coexisting phases much more active than the other, or are active sites maybe only created at the (evolving) phase boundaries? Corresponding answers could be provided by a dedicated analysis of in situ activity data, appropriately accounting for potential flow limitations in the experiment. For the CO oxidation at Pd(100) system such a first analysis of laser-induced fluorescence (LIF) data has in fact already heralded the intriguing contributions this can make [147].

Figure 7.7 shows the LIF-measured CO_2 concentration directly above the catalyst surface, which is a non-invasive local measure of the product formation and therewith of the catalytic activity [147]. Also shown are the corresponding signals as predicted by 1p-kMC–CFD simulations either employing the 1p-kMC lattice model for the pristine Pd(100) surface or the 1p-kMC lattice model for the surface oxide. For the measured range of reaction conditions, namely a temperature ramp at constant pressure and slightly O-rich stoichiometry, only the prior model yields a signature compatible with the experimental data. This suggests the predominant catalytic activity to be due to active sites still being in a metallic surface termination. On the other hand, there is a notable shift of the theoretical signature by $\sim$100 K to lower temperatures. A sensitivity analysis points at the CO oxidation reaction barrier as rate-determining step, and rerunning the simulations on an energetic data base obtained with the less binding RPBE functional [140] indeed brings the theoretical signature into much closer agreement with experiment, cf. Fig. 7.7. In an empirically "correcting" approach one could now attribute the remaining difference to a still deficient RPBE energetics and simply fit the CO oxidation barrier so as to perfectly match the theoretical and experimental LIF profile.

Alternatively, we could recall that the probed reaction conditions fall into the bistability regime discussed above, and both metal and surface oxide phase could

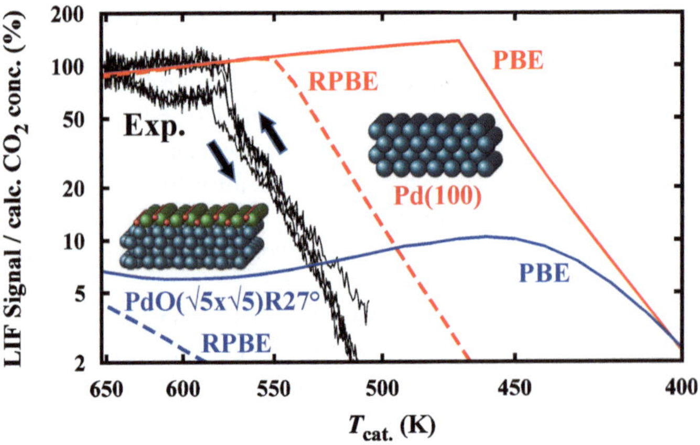

Fig. 7.7 Measured CO_2 LIF signal over the active catalyst surface for a temperature ramp from 500 to 650 K and back (as indicated by the *arrows*). Feed gas conditions are: 4:1 O_2/CO ratio, total pressure 180 mbar; 50% Ar; inlet mass flow 72 ml_n/min. Additionally shown is the corresponding calculated CO_2 concentration variation as predicted for the $(\sqrt{5} \times \sqrt{5})$R27°-O surface oxide (*blue lines*) and for the pristine metal state of Pd(100) (*red lines*). To assess the uncertainties arising from the approximate DFT energetics, data obtained with the PBE [139] (*solid lines*) and RPBE [140] (*dashed lines*) xc functional are shown. From [147]

potentially coexist at the surface [87]. Indeed, at the RPBE level and disregarding any special catalytic activity of sites at domain boundaries, a quantitative agreement with the experimental signature can also be reached when assuming that the predominantly active pristine metal domains cover only a fraction of ~25% of the total surface area [147]. On a methodological level this is a perfect illustration that disagreement of a first-principles theory with experiment can have multiple, quite distinct sources. A naïve fudging of just the 1p energetics to reach agreement in macroscopic observables like catalytic activity is thus ill-advised. Instead, further experiments and/or calculations are required to single out the true source for the disagreement. In the present example, the suggested surface heterogeneity could be scrutinized by combining the LIF activity measurements with an in situ surface characterization technique. If the rationalization in terms of a phase mixture prevails, this could potentially resolve quite some controversies in the emerging field of in situ model catalyst studies. With a prevailing focus on spectro-/microscopic measurements, phases that are predominantly characterized at the working surface have there often tacitly been assumed to also be the ones actuating the catalysis. From a modeling perspective, the truly exciting validation would instead come when multi-lattice or off-lattice kMC simulations are able to explicitly treat an evolving surface heterogeneity. This is the great challenge for the future, and it will for sure create many interesting "ideas" in the context of a reaction-induced dynamical picture of surface catalysis.

7.4 Conclusions and Outlook

Over the last 10–15 years (constrained) ab initio thermodynamics and first-principles microkinetics have become well-established tools in surface catalysis research. (Constrained) ab initio thermodynamics is in fact a routine approach that has spread even well out of academia. It provides first, approximate insight into the structure and composition of the catalyst surface at finite, technologically-relevant gas-phase compositions. Near-term advancement of this technique will most likely center on coupling this thermodynamic framework with global geometry optimization algorithms and thereby overcome the prevalent restricted sampling of configuration space in form of small sets of structural candidates hand-selected by the researcher.

More refined insight into the structure and composition, as well as intrinsic TOFs can be obtained from computationally more involved 1p microkinetic approaches. The power of these techniques and the step-out changes connected with their advent are presently already impressively heralded by trend studies, where rough activity estimates are used for a computational screening of catalyst materials. Such studies are currently based on simplified mean-field kinetic models that assume reaction mechanisms and rate-determining steps, and they employ approximate scaling relations to reduce the required first-principles energetic input. At this level of theory obtaining a detailed mechanistic understanding and quantitative TOFs of an individual system is neither intended, nor achievable. In fact, already the uncertainty in presently-available DFT energetics for extended surface systems will generally prevent reaching quantitative absolute TOFs in the foreseeable future. However, considering that reaching such numbers is similarly elusive in experimental studies this is also not really a goal to worry about. Important is, instead, to systematically validate, e.g. through sensitivity analysis, that the relevant (mechanistic and activity) conclusions drawn are robust with respect to these and other methodological uncertainties.

Due to the continuously increasing computer power alone we will certainly see a rapid spreading of 1p microkinetic modeling in the next years, eventually also into industry. Obvious advancements are the extension to more complex reaction networks with ever diminishing assumptions on reaction paths and intermediates, the move from presently-studied individual facets to entire (supported) nanoparticles, and a gradual shift from prevalent mean-field kinetics to spatially-resolved kinetic Monte Carlo simulations. The central challenge to all of this is that all of the here discussed methodology relies inherently on a rather rigid picture of the catalyst substrate, exploiting a certain level of crystalline order and static active site structures. Addressing the highly dynamic picture of heterogeneous catalysis increasingly suggested by in situ studies—with reaction-induced complex (surface) morphological changes and an evolving, possibly liquid-like phase behavior—is largely impossible with currently available methodology. Great care has to be taken that this incapability to describe such scenarios with present models does not generate the “idea” to readily dismiss them. Instead, it should be a source of motivation to further push the field and tackle the methodological frontiers.

Acknowledgements I am as ever indebted to my research group. Their creativity, curiosity, and enthusiasm for research is my daily motivation. Their diligence, unselfishness, and idealism is the imperative to never lower the pace. I also gratefully acknowledge my temporary second home, the SUNCAT Center at Stanford University. Spending a sabbatical in this stimulating environment has been ideal to assemble the thoughts for this review. This chapter has also been published as Catalysis Letters 146 (3), 541—563 (2016).

References

1. G. Ertl, *Reactions at Solid Surfaces* (Wiley, Hoboken, NJ, 2010)
2. G.A. Somorjai, Y. Li, Proc. Natl. Acad. Sci. U.S.A. **108**, 917–924 (2011)
3. M.A. Newton, Chem. Soc. Rev. **37**, 2644–2657 (2008)
4. R. Schlögl, Angew. Chem. Int. Ed. **54**, 3465–3520 (2015)
5. I. Chorkendorff, H. Niemantsverdriet, *Concepts of Modern Catalysis and Kinetics* (Wiley-VCH, Weinheim, 2003)
6. R.A. van Santen, M. Neurock, *Molecular Heterogeneous Catalysis: A Conceptual and Computational Approach* (Wiley-VCH, Weinheim, 2006)
7. J.K. Nørskov, F. Studt, F. Abild-Pedersen, T. Bligaard, *Fundamental Concepts in Heterogeneous Catalysis* (Wiley, Hoboken, NJ, 2014). ISBN 978-1-118-88895-7
8. A. Stierle, A. Molenbroek, MRS Bull. **32**, 1001–1009 (2007)
9. Ziegler A, Graafsma H, Zhang XF, Frenken JWM (eds) *In-situ Materials Characterization: Across Spatial and Temporal Scales* (Springer, Berlin, 2014)
10. K. Reuter, Oil Gas Sci. Technol. **61**, 471–477 (2006)
11. K. Reuter, Nanometer and sub-nanometer thin oxide films at surfaces of late transition metals. in *Nanocatalysis*, eds. by U. Heiz, U. Landman (Springer, Berlin, 2006). ISBN 978-3-540-32645-8
12. E. Lundgren, A. Mikkelsen, J.N. Andersen, G. Kresse, M. Schmid, P. Varga, Phys.: Condens. Matter **18**, R481–R499 (2006)
13. H. Over, Chem. Rev. **112**, 3356–3426 (2012)
14. J.F. Weaver, Chem. Rev. **113**, 4164–4215 (2013)
15. J. Meyer, K. Reuter, Angew. Chem. Int. Ed. **53**, 4721–4724 (2014)
16. S. Matera, K. Reuter, Catal. Lett. **133**, 156–159 (2009)
17. S. Matera, K. Reuter, Phys. Rev. B **82**, 085446 (2010)
18. S. Matera, M. Maestri, A. Cuoci, K. Reuter, ACS Catal. **4**, 4081–4092 (2015)
19. M.K. Sabbe, M.-F. Reyniers, K. Reuter, Catal. Sci. Technol. **2**, 2010–2024 (2012)
20. K. Reuter, Catal. Lett. **146**, 541–563 (2016)
21. A.F. Voter, Introduction to the Kinetic Monte Carlo Method, in *Radiation Effects in Solids*, ed. by K.E. Sickafus, E.A. Kotomin, B.P. Uberuaga (Springer, Berlin, 2007)
22. K. Reuter, First-Principles Kinetic Monte Carlo Simulations for Heterogeneous Catalysis: Concepts, Status and Frontiers, in *Modelling and Simulation of Heterogeneous Catalytic Reactions: From the Molecular Process to the Technical Syst*em, (Wiley-VCH, Weinhein, 2013). ISBN-10: 3-527-32120-9
23. R.M. Martin, *Electronic-Structure: Basic Theory and Practical Methods* (Cambridge University Press, Cambridge, 2004)
24. K. Burke, J. Chem. Phys. **136**, 150901 (2012)
25. E.A. Carter, Science **321**, 800–803 (2008)
26. P.J. Feibelman, Top. Catal. **53**, 417–422 (2010)
27. C.M. Weinert, M. Scheffler, Mat. Sci. Forum **10–12**, 25–30 (1986)
28. E. Kaxiras, Y. Bar-Yam, J.D. Joannopoulos, K.C. Pandey, Phys. Rev. B **35**, 9625–9635 (1987)

29. M. Scheffler, Thermodynamic Aspects of Bulk and Surface Defects—First-Principles Calculations, in *Physics of Solid Surfaces—1987*, ed. by J. Koukal (Elsevier, Amsterdam, 1987)
30. M. Scheffler, J. Dabrowski, Phil. Mag. A **58**, 107–121 (1988)
31. G.X. Qian, R.M. Martin, D.J. Chadi, Phys. Rev. B **38**, 7649–7663 (1988)
32. X.G. Wang, W. Weiss, S.K. Shaikhutdinov, M. Ritter, M. Petersen, F. Wagner, R. Schlögl, M. Scheffler, Phys. Rev. Lett. **81**, 1038–1041 (1998)
33. X.G. Wang, A. Chaka, M. Scheffler, Phys. Rev. Lett. **84**, 3650–3653 (2000)
34. K. Reuter, M. Scheffler, Phys. Rev. B **65**, 035406 (2001)
35. Z. Lodzianan, J.K. Nørskov, P. Stoltze, J. Chem. Phys. **118**, 11179–11188 (2003)
36. D.A. Mc Quarrie, *Statistical Mechanics* (Harper and Row, New York, 1976)
37. K. Reuter, M. Scheffler, Phys. Rev. B **68**, 045407 (2003)
38. S. Tull, H. Prophet, *JANAF Thermochemical Tables*, 2nd edn. (U.S. National Bureau of Standards, Washington, D.C, 1971)
39. Q. Sun, K. Reuter, M. Scheffler, Phys. Rev. B **67**, 205424 (2003)
40. Q. Sun, K. Reuter, M. Scheffler, Phys. Rev. B **70**, 235402 (2004)
41. D. Loffreda, Surf. Sci. **600**, 2103–2112 (2006)
42. E. Lundgren, J. Gustafson, A. Mikkelsen, J.N. Andersen, A. Stierle, H. Dosch, M. Todorova, J. Rogal, K. Reuter, M. Scheffler, Phys. Rev. Lett. **92**, 046101 (2004)
43. J. Rogal, K. Reuter, M. Scheffler, Phys. Rev. Lett. **98**, 046101 (2007)
44. J. Rogal, K. Reuter, M. Scheffler, Phys. Rev. B **77**, 155410 (2008)
45. G. Zheng, E.I. Altman, Surf. Sci. **504**, 253–270 (2002)
46. S.-L. Chang, P.A. Thiel, J. Chem. Phys. **88**, 2071–2082 (1988)
47. K. Reuter, M. Scheffler, Appl. Phys. A **78**, 793–798 (2004)
48. M. Todorova, E. Lundgren, V. Blum, A. Mikkelsen, S. Gray, M. Borg, J. Gustafson, J. Rogal, K. Reuter, J.N. Andersen, M. Scheffler, Surf. Sci. **541**, 101–112 (2003)
49. P. Kostelník, N. Seriani, G. Kresse, A. Mikkelsen, E. Lundgren, V. Blum, T. Šikola, P. Varga, M. Schmid, Surf. Sci. **601**, 1574–1581 (2007)
50. J. Gustafson, A. Mikkelsen, M. Borg, E. Lundgren, L. Köhler, G. Kresse, M. Schmid, P. Varga, J. Yuhara, X. Torrelles, C. Quiros, J.N. Andersen, Phys. Rev. Lett. **92**,126102 (2004)
51. E. Lundgren, G. Kresse, C. Klein, M. Borg, J.N. Andersen, M. De Santis, Y. Gauthier, C. Konvicka, M. Schmid, P. Varga, Phys. Rev. Lett. **88**, 246103 (2002)
52. A. Michaelides, K. Reuter, M. Scheffler, J. Vac. Sci. Technol., A **23**, 1487–1497 (2005)
53. M.D. Ackermann, T.M. Pedersen, B.L. Hendriksen, O. Robach, S.C. Bobaru, I. Popa, C. Quiros, H. Kim, B. Hammer, S. Ferrer, J.W. Frenken, Phys. Rev. Lett. **95**, 255505 (2005)
54. J. Schnadt, A. Michaelides, J. Knudsen, R.T. Vang, K. Reuter, E. Laegsgaard, M. Scheffler, F. Besenbacher, Phys. Rev. Lett. **96**, 146101 (2006)
55. D.J. Miller, H. Öberg, S. Kaya, H. Sanchez Casalongue, D. Friebel, T. Anniyev, H. Ogasawara, H. Bluhm, L.G.M. Pettersson, A. Nilsson, Phys. Rev. Lett. **107**, 195502 (2011)
56. B. Hendriksen, S. Bobaru, J. Frenken, Surf. Sci. **552**, 229–242 (2004)
57. M. Shipilin, J. Gustafson, C. Zhang, L.R. Merte, A. Stierle, U. Hejral, U. Ruett, O. Gutowski, M. Skoglundh, P.A. Carlsson, E. Lundgren, J. Phys. Chem. C **119**, 15469–15476 (2015)
58. C.T. Campbell, Phys. Rev. Lett. **96**, 066106 (2006)
59. K. Reuter, M. Scheffler, Phys. Rev. Lett. **90**, 046103 (2003)
60. K. Reuter, M. Scheffler, Phys. Rev. B **73**, 045433 (2006)
61. B. Hendriksen, S. Bobaru, J. Frenken, Catal. Today **105**, 234–243 (2005)
62. A. Böttcher, H. Niehus, S. Schwegmann, H. Over, G. Ertl, J. Phys. Chem. B **101**, 11185–11191 (1997)
63. H. Over, Y.D. Kim, A.P. Seitsonen, S. Wendt, E. Lundgren, M. Schmid, P. Varga, A. Morgante, G. Ertl, Science **287**, 1474–1476 (2000)
64. S.J. Roosendahl, A.M. Vredenberg, F.H.P.M. Habraken, Phys. Rev. Lett. **84**, 3366–3369 (2000)

65. R. Kaischew, Commun. Bulg. Acad. Sci. **1**, 100–139 (1950)
66. R. Kaischew, Bull. Acad. Sci. Bulg. **2**, 191–204 (1951)
67. J. Rogal, K. Reuter, M. Scheffler, Phys. Rev. B **69**, 075421 (2004)
68. T. Wang, J. Jelic, D. Rosenthal, K. Reuter, Chem. Cat. Chem. **5**, 3398–3403 (2013)
69. R. Ouyang, J.X. Liu, W.X. Li, J. Am. Chem. Soc. **135**, 1760–1771 (2013)
70. M. Garcia-Mota, M. Rieger, K. Reuter, J. Catal. **321**, 1–6 (2015)
71. C.W. Gardiner, *Handbook of Stochastic Methods* (Springer, Berlin, 2003)
72. G.F. Froment, Catal. Rev. Sci. Eng. **47**, 83–124 (2005)
73. M.J. Hoffmann, S. Matera, K. Reuter, Comp. Phys. Commun. **185**, 2138–2150 (2014)
74. E.W. Hansen, M. Neurock, Chem. Eng. Sci. **54**, 3411–3421 (1999)
75. M. Stamatakis, D.G. Vlachos, J. Chem. Phys. **134**, 214115 (2011)
76. B. Temel, H. Meskine, K. Reuter, M. Scheffler, H. Metiu, J. Chem. Phys. **126**, 204711 (2007)
77. S. Matera, H. Meskine, K. Reuter, J. Chem. Phys. **134**, 064713 (2011)
78. D.J. Liu, A. Garcia, J. Wang, D.M. Ackerman, C.J. Wang, J.W. Evans, Chem. Rev. **115**, 5979–6050 (2015)
79. J.T. Hirvi, T.J.J. Kinnunen, M. Suvanto, T.A. Pakkanen, J.K. Nørskov, J. Chem. Phys. **133**, 084704 (2010)
80. D. Frenkel, B. Smit, *Understanding Molecular Simulation*, 2nd edn. (Academic Press, San Diego, 2002)
81. P. Hänggi, P. Talkner, M. Borkovec, Rev. Mod. Phys. **62**, 251–341 (1990)
82. J.A. Dumesic, D.F. Rudd, L.M. Aparicio, J.E. Rekoske, A.A. Trevin, *The Microkinetics of Heterogeneous Catalysis, American Chemical Society* (1998)
83. S. Müller, J. Phys.: Condens. Matter **15**, R1429–R1500 (2003)
84. Y. Zhang, V. Blum, K. Reuter, Phys. Rev. B **75**, 235406 (2007)
85. M.J. Hoffmann, K. Reuter, Top. Catal. **57**, 159–170 (2014)
86. C. Wu, D. Schmidt, C. Wolverton, W. Schneider, J. Catal. **286**, 88–94 (2012)
87. L.C. Grabow, B. Hvolbaek, J.K. Nørskov, Top. Catal. **53**, 298–310 (2010)
88. A.C. Lausche, A.J. Medford, T.S. Khan, Y. Xu, T. Bligaard, F. Abild-Pedersen, J.K. Nørskov, F. Studt, J. Catal. **307**, 275–282 (2013)
89. G. Henkelman, H. Jonsson, J. Chem. Phys. **115**, 9657–9666 (2001)
90. J.L. Bocquet, Defect diffus. Forum **203**, 81–112 (2002)
91. O. Trushin, A. Karim, A. Kara, T.S. Rahman, Phys. Rev. B **72**, 115401 (2005)
92. F. El-Mellouhi, N. Mousseau, L.J. Lewis, Phys. Rev. B **78**, 153202 (2008)
93. G. Henkelman, G. Johannesson, H. Jonsson, Methods for Finding Saddle Points and Minimum Energy Paths, in *Progress on Theoretical Chemistry and Physics*, ed. by S.D. Schwarz (Kluwer, New York, 2000)
94. H.P. Hratchian, H.B. Schlegel, Finding Minima, Transition States and Following Reaction Pathways on ab initio Potential Energy Surfaces, in *Theory and Applications in Computational Chemistry: The First Forty Years*, ed. by C. Dykstra, G. Frenking, K. Kim, G. Scuseria (Elsevier, Amsterdam, 2005)
95. G. Henkelman, H. Jonsson, J. Chem. Phys. **111**, 7010–7022 (1999)
96. A. Heyden, A.T. Bell, F.J. Keil, J. Chem. Phys. **123**, 224101 (2005)
97. H. Jonsson, G. Mills, K.W. Jacobsen, Nudged Elastic Band Method for Finding Minimum Energy Paths of Transitions, in *Classical and Quantum Dynamics in Condensed Phase Simulations*, ed. by B.J. Berne, G. Cicotti, D.F. Coker (World Scientific, New Jersey, 1998)
98. G. Henkelman, B.P. Uberuaga, H. Jonsson, J. Chem. Phys. **113**, 9901–9904 (2000)
99. E. Weinan, W. Ren, E. Vanden-Eijnden Phys. Rev. B **66,** 052301 (2002)
100. E. Shustorovich, H. Sellers, Surf. Sci. Rep. **31**, 5–119 (1998)
101. E. Hansen, M. Neurock, Surf. Sci. **441**, 410–424 (1999)
102. M. Maestri, K. Reuter, Angew. Chem. Int. Ed. **50**, 1194–1197 (2011)
103. A. Michaelides, Z.P. Liu, C.J. Zhang, A. Alavi, D.A. King, P. Hu, J. Am. Chem. Soc. **125**, 3704–3705 (2003)

104. A. Logadottir, T.H. Rod, J.K. Nørskov, B. Hammer, S. Dahl, C.J.H. Jacobsen, J. Catal. **197**, 229–231 (2001)
105. C.J.H. Jacobsen, S. Dahl, B.S. Clausen, S. Bahn, A. Logadottir, J.K. Nørskov, J. Am. Chem. Soc. **123**, 8404–8405 (2001)
106. H. Toulhoat, P. Raybaud, J. Catal. **216**, 63–72 (2003)
107. S. Linic, J. Jankowiak, M.A. Barteau, J. Catal. **224**, 489–493 (2004)
108. J. Greeley, M. Mavrikakis, Nature Mat. **3**, 810–815 (2004)
109. M.P. Andersson, T. Bligaard, A. Kustov, K.E. Larsen, J. Greeley, T. Johannessen, H. Chris-tensen, J.K. Nørskov, J. Catal. **239**, 501–506 (2006)
110. R.I. Masel, *Principles of Adsorption and Reaction on Solid Surfaces* (Wiley, New York, 1996)
111. R.M. Ziff, E. Gulari, Y. Barshad, Phys. Rev. Lett. **56**, 2553–2556 (1986)
112. K. Reuter, C. Stampfl, M.V. Ganduglia-Pirovano, M. Scheffler, Chem. Phys. Lett. **352**, 311–317 (2002)
113. K. Reuter, M.V. Ganduglia-Pirovano, C. Stampfl, M. Scheffler, Phys. Rev. B **65**, 165403 (2002)
114. F. Abild-Pedersen, J. Greeley, F. Studt, J. Rossmeisl, T.R. Munter, P.G. Moses, E. Skulason, T. Bligaard, J.K. Nørskov, Phys. Rev. Lett. **99**, 016105 (2007)
115. J. Aßmann, D. Crihan, M. Knapp, E. Lundgren, E. Löffler, M. Muhler, V. Narkhede, H. Over, M. Schmid, A.P. Seitsonen, P. Varga, Angew. Chem. Int. Ed. **44**, 917–920 (2005)
116. M.J. Hoffmann, M. Scheffler, K. Reuter, ACS Catal. **5**, 1199–1209 (2015)
117. V.R. Fernandes, J. Gustafson, I.C.H. Svenum, M.H. Farstad, L.E. Walle, S. Blomberg, E. Lundgren, A. Borg, Surf. Sci. **621**, 31–39 (2014)
118. K. Reuter, D. Frenkel, M. Scheffler, Phys. Rev. Lett. **93**, 116105 (2004)
119. M. Rieger, J. Rogal, K. Reuter, Phys. Rev. Lett. **100**, 016105 (2008)
120. K.S. Exner, F. Hess, H. Over, A.P. Seitsonen, Surf. Sci. **640**, 165–180 (2015)
121. M. Boudart, K. Tamaru, Catal. Lett. **9**, 15–22 (1991)
122. C.T. Campbell, Top. Catal. **1**, 353–366 (1994)
123. J.A. Dumesic, J. Catal. **185**, 496–505 (1999)
124. A. Baranski, Solid State Ionics **117**, 123–128 (1999)
125. C. Stegelmann, A. Andreasen, C.T. Campbell, J. Am. Chem. Soc. **131**, 13563 (2009)
126. H. Meskine, S. Matera, M. Scheffler, K. Reuter, H. Metiu, Surf. Sci. **603**, 1724–1730 (2009)
127. M. Kozuch, J.M.L. Martin, Chem. Phys. Chem. **12**, 1413–1418 (2011)
128. D.P. Woodruff, T.A. Delchar, *Modern Techniques of Surface Science* (Cambridge University Press, Cambridge, 1994)
129. J. Wellendorff, T.L. Silbaugh, D. Garcia-Pintos, J.K. Nørskov, T. Bligaard, F. Studt, C.T. Campbell, Surf. Sci. **640**, 36–44 (2015)
130. J.P. Perdew, K. Burke, M. Ernzerhof, Phys. Rev. Lett. **77**, 3865–3868 (1996)
131. J.K. Nørskov, F. Abild-Pedersen, F. Studt, T. Bligaard, Proc. Natl. Acad. Sci. U. S. A. **108**, 937–943 (2011)
132. D. Loffreda, F. Delbecq, F. Vigne, P. Sautet, Angew. Chem. Int. Ed. **48**, 8978–8980 (2009)
133. T. Bligaard, J.K. Nørskov, S. Dahl, J. Matthiesen, C.H. Christensen, J. Sehested, J. Catal. **224**, 206–217 (2004)
134. J.K. Nørskov, T. Bligaard, J. Rossmeisl, C.H. Christensen, Nature Chem. **1**, 37–46 (2009)
135. F. Besenbacher, I. Chorkendorff, B.S. Clausen, B. Hammer, A.M. Molenbroek, J.K. Nørskov, I. Stensgaard, Science **279**, 1913–1915 (1998)
136. B. Hammer, L.B. Hansen, J.K. Nørskov, Phys. Rev. B **59**, 7413–7421 (1999)
137. J.J. Mortensen, K. Kaasbjerg, S.L. Frederiksen, J.K. Nørskov, J.P. Sethna, K.W. Jacobsen, Phys. Rev. Lett. **95**, 216401 (2005)
138. J. Wellendorff, K.T. Lundgaard, A. Møgelhøj, V. Petzold, D.D. Landis, J.K. Nørskov, T. Bligaard, K.W. Jacobsen, Phys. Rev. B **85**, 235149 (2012)
139. B.L.M. Hendriksen, M.D. Ackermann, R. van Rijn, D. Stoltz, I. Popa, O. Balmes, A. Resta, D. Wermeille, R. Felici, S. Ferrer, J.W.M. Frenken, Nature Chem. **2**, 730–734 (2010)
140. A.T. Fromhold, *Theory of Metal Oxidation*, vols. I, II, (North-Holland, Amsterdam, 1976)

141. R.A. van Santen, M. Neurock, S.G. Shetty, Chem. Rev. **110**, 2005–2048 (2010)
142. J.K. Nørskov, T. Bligaard, A. Logadottir, S. Bahn, L.B. Hansen, M. Bollinger, H. Bengaard, B. Hammer, Z. Sljivancanin, M. Mavrikakis, Y. Xu, S. Dahl, C.J.H. Jacobsen, J. Catal. **209**, 275–279 (2002)
143. J. Wellendorff, K.T. Lundgaard, K.W. Jacobsen, T. Bligaard, J. Chem. Phys. **140**, 144107 (2014)
144. A.J. Medford, J. Wellendorff, A. Vojvodic, F. Studt, F. Abild-Pedersen, K.W. Jacobsen, T. Bligaard, J.K. Nørskov, Science **345**, 197–200 (2014)
145. S. Blomberg, M.J. Hoffmann, J. Gustafson, N.M. Martin, V.R. Fernandes, A. Borg, Z. Liu, R. Chang, S. Matera, K. Reuter, E. Lundgren, Phys. Rev. Lett. **110**, 117601 (2013)
146. V.M. Janardhanan, O. Deutschmann Computational Fluid Dynamics of Catalytic Reactors, in *Modelling and Simulation of Heterogeneous Catalytic Reactions: From the Molecular Process to the Technical System*, ed. by O. Deutschmann (Wiley-VCH, Weinheim, 2013). ISBN-10: 3-527-32120-9
147. S. Matera, S. Blomberg, M.J. Hoffmann, J. Zetterberg, J. Gustafson, E. Lundgren, K. Reuter, ACS Catal. **5**, 4514–4518 (2015)

Chapter 8
Catalysis Engineering: From the Catalytic Material to the Catalytic Reactor

Stefano Rebughini, Mauro Bracconi, Alberto Cuoci and Matteo Maestri

Abstract This chapter deals with the application of chemical reaction engineering and computational fluid dynamics (CFD) for the analysis and assessment of the interactions between mass and heat transport and chemical reactions. In the first part of the Chapter, we review fundamental concepts of chemical reaction engineering, by showing the potential impact of transport phenomena at the macroscale on the observed functionality of the catalytic material. This includes both the effect of the distribution of the residence times in the reactor and the impact of internal and external transport phenomena. In the second part, we illustrate modern approaches to catalytic reaction engineering based on CFD simulations. In particular, we present the algorithms to couple microkinetic models and kinetic Monte Carlo (kMC) simulations with CFD. The potentialities of the method are assessed by means of a showcase of the CFD-based analysis of a spectroscopic cell for *operando* experiments. This example clearly shows that transport artifacts in standard equipment may lead to an erroneous interpretation of the experiments if not properly accounted for.

Book chapter to appear in.

"*Operando Studies in Heterogeneous Catalysis*".

Edited by Irene M.N. Groot and Joost W.M. Frenken, Leiden University, The Netherlands.

Springer.

S. Rebughini · M. Bracconi · M. Maestri (✉)
Laboratory of Catalysis and Catalytic Processes - Dipartimento di Energia, Politecnico di Milano, Via La Masa 34, 20156 Milan, Italy
e-mail: matteo.maestri@polimi.it

A. Cuoci
Dipartimento di Chimica, Materiali e Ingegneria Chimica "G. Natta", Politecnico di Milano, Piazza Leonardo da Vinci 32, 20133 Milan, Italy

J. Frenken and I. Groot (eds.), Operando *Research in Heterogeneous Catalysis*,
Springer Series in Chemical Physics 114, DOI 10.1007/978-3-319-44439-0_8

8.1 Introduction

Catalysts are functional materials or molecules able to provide ‘active sites’ which allow for a stabilization of the Gibbs free energy of the transition state of the elementary reactions, resulting in an enhanced reaction rate. In heterogeneous gas-solid catalysis, such ‘active sites’ are, by nature, present in a separate solid phase. Thus, the reactants in the gas-phase have to reach the ‘active sites’ and this involves several physical transport processes [1]. These phenomena are characterized by a wide range of characteristic time and length scales, which can cover many orders of magnitudes, as shown in Fig. 8.1. At the microscale, the electronic interaction between the active sites and the reactants drives the capability of the catalyst in cleaving/forming specific bonds, which results in different atomistic events (elementary steps) at the catalyst surface. The interplay between the rates of different steps along with coverage and morphological effects (e.g. different active sites and their relative distribution on the surfaces) is the key event at the mesoscale. Here, the different rates of the elementary steps result in different catalytic cycles and reaction mechanisms, with a strong impact on the observed performances of the material. Finally, transport properties from the bulk fluid flow to the catalyst interface are characterized by longer time and length scales and represent the macroscale. All these steps occur at a finite rate determining that concentration and temperature gradients may arise between the bulk fluid phase and the active sites. Thus, the physical steps may have a strong impact on the rate of the overall process and, as a consequence, the observable reaction rate may differ substantially from the intrinsic reaction rate of the chemical transformation under bulk fluid-phase conditions. Therefore, it is of crucial importance to distinguish between the *intrinsic functionality* of the catalyst material and its *macroscopically observed behavior*. The former is related to the presence of the ‘active sites’ and their ability of making/breaking the chemical bonds when interacting with molecules and it is determined primarily by the electronic structure at the atomic scale. On the

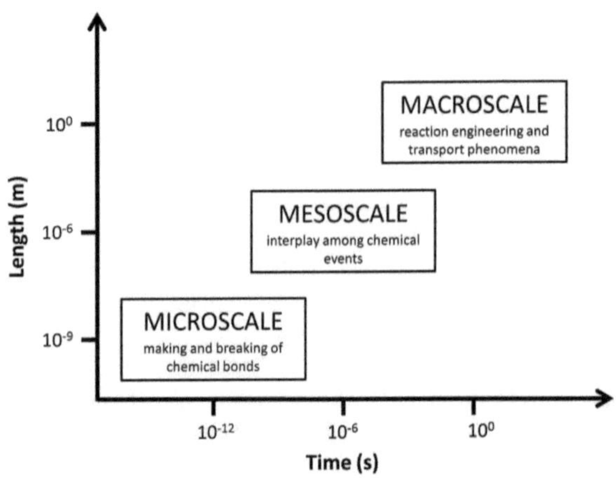

Fig. 8.1 Sketch of different scales involved in a chemical process

contrary, the latter is the actual observed behavior of the catalyst in the chemical reactor and it is the result of the interplay *among all the phenomena at the different scales*. Thus, the *macroscopically observed behavior* is intrinsically a *multiscale property of the system* and its analysis and design require to properly link simulations and models at the different scales [2]. Typically, at the microscale the calculation of the kinetic parameters of the elementary reactions is performed within the validity of the harmonic Transition State Theory (h-TST). In this context, the most important information for the determination of a rate constant is the location of the transition state, i.e. the calculation of the activation energy. Electronic structure theories explicitly treat the electronic degrees of freedom and are the natural basis for the calculation of the potential energy surface. In particular, for the extended metal surfaces typical in catalytic problems, Density Functional Theory (DFT) is currently the most adopted method among the electronic-structure theories [3]. Moreover, in view of the huge computational cost connected to such calculations, semi-empirical and less demanding methods are also used, especially for the exploration of complex reaction networks [4].

At the mesoscale, the interplay between the rates of the different elementary steps is accounted for. This interplay is crucial in determining the dominant paths, which take part in the catalytic cycle at the given operating conditions. Statistical simulations are used to explicitly account for the interplay between all the chemical events. In particular, master-equation-based kinetic Monte Carlo (kMC) algorithms are employed in order to account for the correct and site-resolved statistical interplay among the elementary steps of the reaction network [3]. A special case is represented by the “mean-field” approximation, which relies on the assumption of very fast diffusion of the species at the surface [5].

At the macroscale, transport phenomena are responsible for the macroscopic distribution of the velocity of the fluid (hydrodynamics), temperature, and composition in the reactor. At the length of the characteristic dispersion of the reactor (e.g. solid particles), they are responsible for the mixing, temperature, and concentrations at the interface between the phases. The modeling of the reactor device is based on either indefinite or macroscopic balance equations for the conservation of mass, energy, and momentum. The behavior of the macroscale is mainly determined by fluid-dynamics. By increasing the complexity and (in principle) the accuracy of the model, one can move to two-dimensional (2D) or three-dimensional (3D) macroscopic models up to the computational solution of the Navier-Stokes equations for the real reactor geometry by means of Computational Fluid Dynamics (CFD).

This chapter introduces the application of chemical reaction engineering and CFD to obtain a better understanding of the interactions between mass and heat transport and chemical reactions in catalytic reactors. We mainly focus on the difference between the intrinsic behavior of the catalytic material and its observed behavior in reacting conditions (catalytic reactor). First, we review some basic concepts of chemical reaction engineering, in order to clearly demonstrate the impact of transport phenomena at the macroscale on the observed functionality of the material. This comprises both the effect of the distribution of the residence times in the reactor and the impact of internal and external transport phenomena. Then we show modern approaches to catalytic reaction engineering based on CFD simulations. In particular, we present the algorithms to couple microkinetic models and kMC simulations with CFD. To conclude, we report as a showcase the CFD-based analysis of a spectroscopic cell for *operando* experiments.

8.2 Intrinsic Versus Observed Reaction Rate

The chemical reactor is the device where the physicochemical transformations are caused [6]. There are a huge number of different reactor types (e.g. fixed bed reactors, fluidized bed reactors, stirred tank reactors, structured reactors) and the geometry of the reactor is specifically conceived and designed in order to ensure a proper interaction between the catalyst and the reactants. In this section, we provide an overview of the main factors at the macroscale, which may strongly affect the observed reaction rate in the reactor. In particular, the geometry and the operating conditions of the chemical reactor affect two crucial aspects:

(a) the distribution of the residence time in the reactor;
(b) the mechanisms and the rates of transport properties.

Both these features may strongly influence the observed functionality of the catalytic material.

8.2.1 Residence Time Distribution (RTD)

The Residence Time Distribution (RTD) can be defined as the information how long an infinitesimal volume of fluid resides in the reactor [6]. Assuming a Lagrangian perspective, the residence time of a fluid element is defined as the time elapsed since it entered the reactor. Depending on the mixing rate characterizing the device, we may experience that a portion of the reactant fluid is held in the reactor for a longer time, whereas another portion is held for a shorter time. As such, the hydrodynamics in the reactor strongly influence the time spent by the molecules in the reactor. To illustrate this issue, let us consider a simple experiment, where at a given time t_0, a pulse of a fluid tracer is fed to a reactor and the composition of the outlet stream is being monitored. As shown in Fig. 8.2, we can identify two asymptotic behaviors. In case of fully segregated flow (i.e. no mixing), all the molecules of the tracer will experience the same residence time. Therefore, no marked elements will be observed at the outlet of the reactor until the residence time of the reactor is elapsed. In case of very fast and perfect mixing, instead, at the injection time all the volume of the reactor will reach a uniform concentration of the tracer. The elements of fluid leaving the reactor will reflect such instantaneous mixing of the tracer at the injection time, thus resulting in an exponential decay of the initial concentration in response to the inflow of the unmarked fluid. In case of

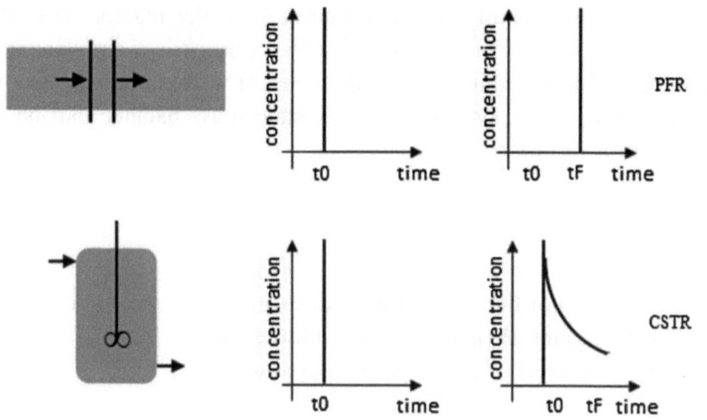

Fig. 8.2 Concentration of tracer species using an impulse input in a PFR (*top*) and in a CSTR (*bottom*)

intermediate level of mixing, a portion of the fluid will exit at a time less than the mean residence time, another portion at the mean residence time, and the remaining part will stay in the reactor for a time longer than the mean residence time.

The two asymptotic behaviors of fully mixed and fully segregated flow represent ideal and extreme situations of RTD and can be easily modeled in terms of macroscopic conservation equations. These two extreme situations are usually referred to as "ideal reactor models" and they are of primary importance in the interpretation and analysis of experimental data.

8.2.1.1 Fully Mixed Reactor: Continuous Stirred Tank Reactor (CSTR)

The Continuous Stirred Tank Reactor is assumed to be perfectly mixed. Due to the very fast mixing, the composition of each species is homogeneous in all the portions of the reactor volume. Thus, the chemical reactions occur at uniform operating conditions throughout all the reactor volume. Under these assumptions, the steady-state macroscopic mass balance for the generic *i*-th species can be written as:

$$0 = F_i^{IN} - F_i + R_i V \tag{8.1}$$

where F_i is the stream of the *i*-th component, expressed in mole per unit of time, R_i is the reaction term, and V is the whole reactor volume.

8.2.1.2 Fully Segregated Reactor: Plug Flow Reactor (PFR)

In case of absence of mixing, each volume of the fluid entering the reactor will travel along the reactor volume without any interaction with the other parts of the reactor volume. Consequently, the concentration in the reactor will be different along the reactor volume. An example of such a reactor is a circular pipe with an axial length much longer than the cross diameter. If we assume a perfect mixing in the radial direction, the macroscopic steady-state mass balance can be written in differential form as follows:

$$\frac{dF_i}{dV} = R_i \tag{8.2}$$

where F_i is the stream of the i-th component, expressed in mole per unit of time, R_i is the reaction term, and dV is the generic infinitesimal volume. At the inlet of the reactor, the concentration of each species is known:

$$F_i(V = 0) = F_i^0 \tag{8.3}$$

To illustrate the effect of mixing on the observed conversion, let us consider the following simple example. We assume to perform the same reaction

$$A \rightarrow B \tag{8.4}$$

in two isothermal and isobaric reactors. A and B are two generic species and their consumption and production rates are described by a global first-order reaction rate:

$$r = kC_A \tag{8.5}$$

where k is the kinetic constant, expressed in s^{-1}, and C_A is the reactant concentration, expressed in mole per unit of volume.

The two reactors have the same volume and all the operating conditions are assumed identical. The only difference is that the first reactor operates with no mixing (identical to a PFR reactor) and the second one is fully mixed (identical to a CSTR reactor). The concentration of A at the end of the two reactors can be derived analytically by the previous equations. In particular, for the CSTR reactor the outlet concentration of A turns out to be:

$$C_A = \frac{C_A^0}{1 + k\tau} \tag{8.6}$$

whereas the integration of (8.2) and (8.3) for the PFR reactor leads to:

$$C_A = C_A^0 \exp(-k\tau) \tag{8.7}$$

where τ is the residence time. The two outlet concentrations are different and, in particular, we observe that the PFR reactor shows a higher conversion than the CSTR reactor for the same operating conditions. This different observed behavior depends only on the different distribution of the residence time in the reactors.

8.2.2 Interaction Between Chemical Rates and Transport Rates

In heterogeneous catalytic systems, the active sites are by nature present in a different phase with respect to the flow of the reactants. Figure 8.3 shows a schematic representation of the main physical and chemical steps involved in the catalytic process inside a reactor: (1) inter-phase diffusion of the reactants from the bulk to the catalyst surface through the boundary layer; (2) intra-phase diffusion of the reactants through the porous structures of the catalyst; (3–5) adsorption, surface reaction, and desorption steps; (6) intra-phase counter-diffusion of the products by the pores; and (7) intra-phase counter diffusion of the products from the catalyst surface to the bulk of the gas phase through the boundary layer. All these steps occur at a finite rate and thus concentration (and temperature) gradients may arise between the bulk fluid phase and the proximity of the active site. Therefore, depending on the relative velocity of physical and chemical processes, the observable reaction rate may differ substantially from the intrinsic reaction rate of the chemical transformation under bulk fluid-phase conditions. In this paragraph, a simplified description of the effect of the physical steps of Fig. 8.3 on the observable reaction rates is described and discussed. First, we will analyze the effects of the transport from the bulk to the catalyst surface. Then, we will show how the transport inside the catalyst may substantially affect the observed reaction rate.

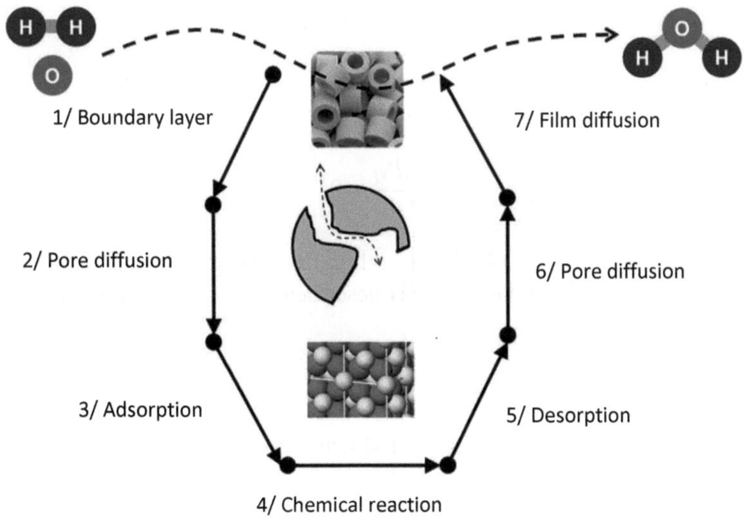

Fig. 8.3 Main physical and chemical steps involved in the catalytic process

8.2.3 External Transport Effects

The physical phenomena of transport from the bulk of the gas phase to the catalyst surface are described by steps (1) and (7) of Fig. 8.3, which are purely diffusive and follow the Stefan-Maxwell equations:

$$\nabla x_i = \sum_{\substack{j=1 \\ j \neq i}}^{NS} \frac{1}{C_{tot}\Gamma_{i,j}}(x_i N_j - x_j N_i) \tag{8.8}$$

where x_i is the mole fraction of the i-th component, C_{tot} is the total concentration, N_i is the molar flux of the i-th component, and $\Gamma_{i,j}$ is the binary diffusion coefficient between the i-th and the j-th components. This equation in the case of a two-component mixture at constant pressure and with equimolar counter-diffusion is reduced to the First Fick's Law:

$$N_i = -\Gamma_{i,j}\nabla C_i \tag{8.9}$$

The flux of the i-th component is obtained by solving (8.9) with the proper boundary conditions. If the thickness of the boundary layer δ is small compared to the curvature of the catalyst, the problem reduces to:

$$N_i = -\Gamma_{i,j}\frac{dC_i}{dx} \tag{8.10}$$

Due to the conservation of mass, the flux through the stagnant film must be constant, thus the derivative of the flux is zero:

$$\frac{dN_i}{dx} = 0 \tag{8.11}$$

By combining (8.10) and (8.11) and by assuming constant diffusivity, the transport through the boundary layer is described by the following equation:

$$\frac{d^2 C_i}{dx^2} = 0 \tag{8.12}$$

The solution of (8.12) requires the following boundary conditions:

$$\begin{cases} C_i(x=0) = C_i^S \\ C_i(x=\delta) = C_i^B \end{cases} \tag{8.13}$$

The second boundary condition is based on the assumption that the fluid is well mixed, thus the concentration of the i-th component at the edge of the stagnant film is equivalent to the concentration in the bulk. The solution of (8.12) leads to:

$$C_i = C_i^S + (C_i^B - C_i^S)\frac{x}{\delta} \quad (8.14)$$

and the molar flux is:

$$N_i = -\frac{\Gamma_{i,j}}{\delta}(C_i^B - C_i^S) \quad (8.15)$$

However, the thickness δ of the boundary layer is unknown. Thus the molar flux is written in terms of a mass-transfer coefficient k_C:

$$N_i = k_C(C_i^B - C_i^S) \quad (8.16)$$

In the case of a generic first-order kinetics $A \to B$, where the kinetic constant is k_S, the flux of A is equal to the reaction rate to prevent the accumulation or depletion at steady state conditions:

$$r = k_S C_A^S = k_C(C_A^B - C_A^S) \quad (8.17)$$

By estimating C_A^S from (8.17), the rate expression in terms of the measurable concentration in the bulk, C_A^B, is obtained:

$$r = \frac{1}{\frac{1}{k_S} + \frac{1}{k_C}} C_A^B \quad (8.18)$$

An observed reaction rate constant can be defined in terms of k_S and k_C:

$$\frac{1}{k_{OBS}} = \frac{1}{k_S} + \frac{1}{k_C} \quad (8.19)$$

A reaction rate expressed in terms of observable quantities is simply given by:

$$r_{OBS} = k_{OBS} C_A^B \quad (8.20)$$

and the Damkholer dimensionless number can be defined as:

$$Da = \frac{k_S L}{k_C} \quad (8.21)$$

where L is the characteristic length of the catalyst geometry (e.g. the diameter for spherical catalysts). The Damkholer number and the observable kinetic constant can be used to investigate the influence of diffusional resistance on the observed reaction rate in the reactor. In particular, two asymptotic conditions can be identified:

1. *Chemical regime*: this regime occurs when the reaction is slower than the transport in the boundary layer. In this case the kinetic constant is smaller than the mass-transfer coefficient ($k_S \ll k_C$), thus the observed reaction rate is:

$$\begin{aligned} r_{OBS} &= k_S C_i^B \\ k_{OBS} &\rightarrow k_S \end{aligned} \tag{8.22}$$

   This results in a very small Damkholer number ($Da \rightarrow 0$). Thus, the observed reaction rate is controlled only by the reaction that occurs at the catalyst. In these conditions, the intrinsic and the observed reaction rates are identical.
2. *Fully external mass-transfer regime*: this regime appears for very fast surface reactions. In this case, the kinetic constant is higher than the mass-transfer coefficient ($k_S \gg k_C$) and the Damkholer number is high ($Da \rightarrow \infty$). In these conditions, the observed reaction rate turns out to be a first-order reaction (irrespective of the kinetic order of the rate equation) and the kinetic constant is equal to the mass-transfer coefficient:

$$\begin{aligned} r_{OBS} &= k_C C_i^B \\ k_{OBS} &\rightarrow k_C \end{aligned} \tag{8.23}$$

   Thus, in these conditions, external diffusion fully masks the intrinsic kinetics specific of the catalyst material.

8.2.4 Internal Transport Effects

Many solid catalysts contain pores in order to increase the specific surface area available for adsorption and reaction. The 'active sites' are thus located in the pore network and the diffusion of molecules in confined spaces (steps (2) and (6) of Fig. 8.3) can become limiting as well. This phenomenon of diffusion of gas species inside a pore structure is described by the Knudsen diffusion, which assumes that the molecules interact more often with the pore walls than with other molecules. However, in order to properly describe the diffusion inside the catalyst pores, also the molecule-molecule collision has to be considered. Therefore, only with a coupling of the molecular diffusion with the Knudsen diffusion, the transport phenomena in a porous catalyst can be adequately described. In particular, in case of equimolar counter-diffusion of a binary mixture, the transition diffusivity Γ_i^T of the *i*-th component, can be approximated by the Bosanquet equation:

$$\frac{1}{\Gamma_i^T} = \frac{1}{\Gamma_i^B} + \frac{1}{\Gamma_i^K} \tag{8.24}$$

where Γ_i^B is the molecular diffusion and Γ_i^K is the Knudsen diffusion.

Now, we consider the idealized situation of a cylindrical pore in a catalytic slab. For an isothermal, isobaric, first-order reaction that occurs on the pore wall $A \rightarrow B$, the mole balance on a slice of Δx thickness can be written as:

$$\pi R_{pore}^2 N_A|_x - \pi R_{pore}^2 N_A|_{x+\Delta x} - k_s C_A^S (2\pi R_{pore})\Delta x = 0 \tag{8.25}$$

where N_A is the molar flux of A estimated at both sides of the slice, k_S is the kinetic constant, expressed in terms of catalytic area and $(2\pi R_{pore})\Delta x$ is the area of the pore wall in the catalyst slice. By changing (8.25) and taking the limit as $\Delta x \rightarrow 0$, the mole balance can be re-written as:

$$-\frac{dN_A}{dx} = \frac{2k_S}{R_{pore}} C_A \tag{8.26}$$

Fick's Law (8.9) relates the molar fluxes to the concentration gradients and its substitution in (8.26) yields the following second-order differential equation:

$$\frac{d^2 C_A}{dx^2} - \frac{2k_S}{\Gamma_A^T R_{pore}} C_A = 0 \tag{8.27}$$

where Γ_A^T is assumed constant. The surface rate constant can be re-written on a volume basis by using the surface-to-volume ratio:

$$\frac{A}{V} = \frac{2}{R_{pore}} \tag{8.28}$$

and the mole balance in terms of kinetic constant on volume basis k_S':

$$\frac{d^2 C_A}{dx^2} - \frac{k_S'}{\Gamma_A^T} C_A = 0 \tag{8.29}$$

Equation (8.29) can be simplified by considering the dimensionless length:

$$x^* = \frac{x}{L} \tag{8.30}$$

and the generalized Thiele modulus:

$$\phi = L\sqrt{\frac{k'_S}{\Gamma_A^T}} \tag{8.31}$$

where L is the characteristic length of the catalyst geometry, which for general geometries is taken as the volume-to-surface ratio. The substitution of (8.30) and (8.31) in (8.29) leads to:

$$\frac{d^2 C_A}{dx^{*2}} - \phi^2 C_A = 0 \tag{8.32}$$

The boundary conditions required for the solution of this differential equation are:

$$\begin{cases} C_A(x^* = 0) = C_A^S \\ \dfrac{dC_A}{dx^*}(x^* = 1) = 0 \end{cases} \tag{8.33}$$

The analytical solution of (8.32) gives:

$$C_A = C_A^S \frac{\cosh[\phi(1 - x^*)]}{\cosh[\phi]} \tag{8.34}$$

Figure 8.4 shows that—depending on the values of the Thiele modulus—concentration gradients can arise in the catalytic slab, thus affecting the overall reaction rate. In particular, when the Thiele modulus is small, transport is faster than reaction and consequently the profile is flat. On the other hand, when the Thiele modulus is large, a significant diffusional resistance prevents a constant concentration profile inside the catalyst. This effect of the intra-porous diffusion on the observed reaction rate is usually expressed in terms of the effectiveness factor η, defined as:

$$\eta = \frac{r_{OBS}}{r_{MAX}} = \frac{\int_0^V r(C_A)dV}{r(C_A)V} \tag{8.35}$$

where V is the catalyst volume. The effectiveness factor is a useful parameter to quantify and represent the *internal mass-transfer regime*. When the concentration profile is flat, it is equal to one and all the active sites work at the same conditions of concentrations. When concentration gradients establish, the reaction rate is not uniform in the solid and thus the efficiency is lower than 1. Integrating (8.35) over the catalyst volume for first-order kinetics leads to the generic definition of the effectiveness factor as a function of the Thiele modulus:

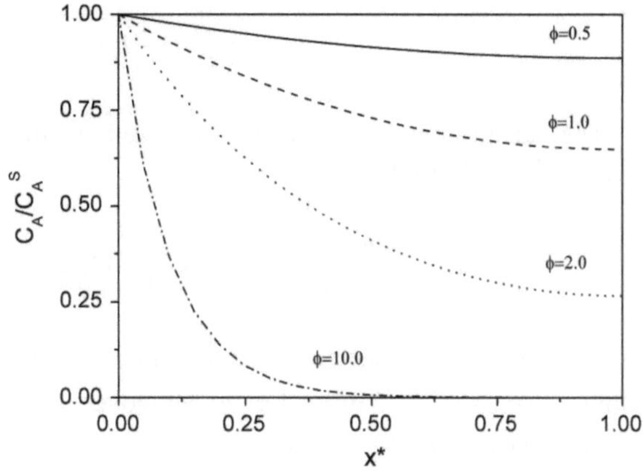

Fig. 8.4 Effect of the Thiele modulus on the normalized concentration profile in a catalyst pore with generic first-order surface reaction

$$\eta = \frac{\tanh(\phi)}{\phi} \tag{8.36}$$

It can be demonstrated that for sufficiently high values of the Thiele modulus, the effectiveness factor has the following functional form, irrespective of the specific geometry of the solid:

$$\eta = \frac{1}{\phi} \tag{8.37}$$

The generalized Thiele modulus, defined in (8.31), can be extended to more complex reactions. For instance, the generalized Thiele modulus for an irreversible reaction of order n is:

$$\phi = \frac{V}{A}\sqrt{\frac{n+1}{2}\frac{k_S C_{A,S}^{n-1}}{\Gamma_A^T}} \qquad n > -1 \tag{8.38}$$

However, despite the larger complexity of the Thiele-modulus definition, the asymptotic value of the effectiveness factor (at large values of the Thiele modulus) is still $\frac{1}{\phi}$.

We now aim to quantify the effect of such transport limitations on the observable parameters. In this scope, the observed reaction rate can be written in terms of the intrinsic rate expression and the effectiveness factor:

$$r_{OBS} = \eta k_S C_{A,S}^n \tag{8.39}$$

and, assuming a large value of the Thiele modulus, we can write:

$$r_{OBS} = \frac{1}{\phi} k_S C_{A,S}^n \tag{8.40}$$

The substitution of (8.37) in (8.40) leads to the following expression for the observed reaction rate:

$$r_{OBS} = \frac{A}{V} \left(\frac{2}{n+1} \Gamma_A^T k_S \right)^{1/2} C_{A,S}^{(n+1)/2} = k_{OBS} C_A^{n_{OBS}} \tag{8.41}$$

The order of the reaction observed under conditions of severe diffusional limitations is $n_{OBS} = \frac{(n+1)}{2}$ instead of n. Moreover, if an Arrhenius expression for the kinetic constant is considered:

$$k = k_0 \exp\left(-\frac{E}{RT} \right) \tag{8.42}$$

the activation energy for the observed kinetic constant is:

$$E_{OBS} = \frac{E_S + E_\Gamma}{2} \tag{8.43}$$

where E_S is the true activation energy for the reaction and E_Γ is the activation energy for the diffusion. Diffusional processes, however, are weakly activated compared to the chemical reactions, thus the contribution of E_Γ can be considered negligible and the observed activation energy for a severely diffusion-limited process can be estimated half of the true value.

8.2.5 Temperature Dependence of the Observed Reaction Rate

Let us summarize the different regimes that we have identified in the analysis as a function of temperature. Figure 8.5 shows the Arrhenius plot for the observed activation energy at different temperatures. At low temperature, kinetics is slower than transport and thus it is the controlling step. Thus, the activation energy is not affected by the presence of the diffusion. By increasing the temperature, gradients start to establish in the solid and the efficiency drops to values lower than 1. According to (8.43), the observed activation energy becomes half that of the intrinsic activation energy. At very high temperature, reaction is so fast that the system is under external mass-transfer limitations. As shown in Sect. 8.2.3, in these conditions the observed reaction constant is only dominated by transport, which is not an activated process. Therefore, the apparent activation energies approach very small values.

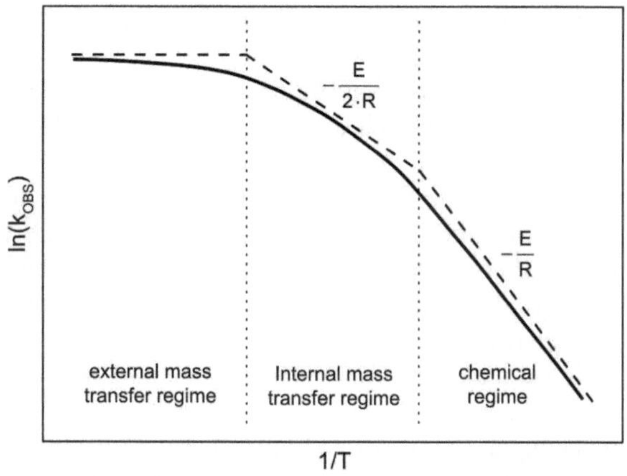

Fig. 8.5 Temperature dependence of the observed rate constant of a reaction occurring in a porous catalyst

Overall, in this section, we have shown by means of simple and basic concepts of chemical reaction engineering how the physical steps which are required for interaction between the active sites and the reactants can considerably affect the observed reaction rates. These issues have to be always clearly kept in mind both during the design of the equipment and during the interpretation of experiments.

8.3 Computational Fluid Dynamics of Gas-Solid Catalytic Reactors

8.3.1 Multiscale Modeling

In the previous section, we have shown how phenomena at the macroscale can strongly affect the observed macroscopic functionality of a catalyst material by means of simplified approaches. Here, we concentrate on the description of the methodologies that have been recently proposed to model in detail the interaction between the macroscale and the micro- and the mesoscales [2]. On one side, the most natural approach would be to proceed according to a bottom-up philosophy, from the atomic scale to the reactor scale. In particular, a fully "first-principles" approach (i.e. *by solving the fundamental governing equations at each scale*) would be desirable to allow for a very detailed and fundamental analysis of the prediction in view of a rational design of the observed functionality. However, this approach is hampered by several issues, which mainly concern with the coupling of very different time scales. In fact, the resolution of the macroscopic scale with the small time step required for the microscale would lead to an enormous waste of computational time, since the continuum dynamics usually evolves on a much larger time scale. Therefore, bottom-up approaches based on direct coupling, even if desirable (especially in view of a detailed and rational understanding of the phenomena) *are not of practical relevance* and their application is mainly related to very simple and ideal cases, where the focus is more on the method development rather than on the application [7].

On the contrary, most of the methods proposed and applied for the simulation of gas-solid catalytic reactors are based *on the effective decoupling of the interdependencies among the scales*. The main and general feature of these methods is *to treat each scale separately*, either by using specific numerical procedure or by introducing specific approximations in order to effectively decouple the time scales of the different scales. In doing so, each scale is simulated with different time steps, thus avoiding the direct coupling of the phenomena occurring at very different characteristic times.

In particular, our focus is on the coupling between the computational solution of the transport equations at the macroscale with kMC simulations and microkinetic modeling at the meso- and microscales, respectively.

The governing equations at the macroscale are summarized in the following.

8.3.2 Governing Equations

8.3.2.1 Gas Phase

The reactive flows under investigation in the present work are mathematically described by the conservation equations for continuous, multicomponent, compressible, thermally-perfect mixtures of gases [8]. The conservation equations of total mass, momentum, individual-species mass fractions and energy for a Newtonian fluid are reported in the following:

$$\frac{\partial \rho}{\partial t} + \nabla \cdot (\rho \mathbf{v}) = 0 \tag{8.44}$$

$$\frac{\partial}{\partial t}(\rho \mathbf{v}) + \nabla \cdot (\rho \mathbf{v} \otimes \mathbf{v}) = -\nabla \cdot (p\mathbf{I}) + \nabla \cdot \left[\mu\left(\nabla \mathbf{v} + (\nabla \mathbf{v})^T\right) - \frac{2}{3}\mu(\nabla \cdot \mathbf{v})\mathbf{I} \right] + \rho \mathbf{g} \tag{8.45}$$

$$\frac{\partial}{\partial t}(\rho \omega_k) + \nabla \cdot (\rho \omega_k \mathbf{v}) = -\nabla \cdot (\rho \omega_k \mathbf{V}_k) \qquad k = 1, \ldots, NCG \tag{8.46}$$

$$\rho \hat{C}_P \frac{\partial T}{\partial t} + \rho \hat{C}_P \mathbf{v} \cdot \nabla T = \nabla \cdot (\lambda \nabla T) - \rho \nabla T \cdot \sum_{k=1}^{NCG} \hat{C}_{P,k} \omega_k \mathbf{V}_k \tag{8.47}$$

In the equations reported above, t is the time, p and T are the pressure and the temperature, respectively, ρ the density, λ the thermal conductivity, μ the dynamic viscosity, and $\hat{C}_P$ the specific heat at constant pressure of the gas phase mixture, $\mathbf{v}$ the velocity vector, $\mathbf{g}$ the acceleration vector due to the gravity, and $\mathbf{I}$ the identity tensor. NGC is the number of gas-phase species. The subscript k refers to the individual gas-phase species and ω_k is the corresponding mass fraction, $\mathbf{V}_k$ is the diffusion velocity, and $\hat{C}_{P,k}$ the specific heat at constant pressure.

The density of the mixture is calculated using the equation of state of ideal gases. Both Fickian and thermal diffusion are taken into account for evaluating the diffusion velocities, according to the approach suggested in [9]:

$$\mathbf{V}_k = -\frac{\Gamma_{k,mix}}{\omega_k}\nabla\omega_k - \frac{\Gamma_{k,mix}\vartheta_k}{X_k}\frac{1}{T}\nabla T \tag{8.48}$$

where $\Gamma_{k,mix}$ is the mixture diffusion coefficient, X_k the mole fraction, and ϑ_k the thermal diffusion ratio of species k-th. Mass conservation is enforced by employing the approach proposed by [10], based on the definition of a conservation diffusion velocity. In particular, in this approach the corrected diffusion velocity vector $\mathbf{V}_k^C$ (to be used in (8.46) and (8.47)) is given as:

$$\mathbf{V}_k^C = \mathbf{V}_k + \mathbf{V}_C \tag{8.49}$$

where $\mathbf{V}_C$ is a constant correction factor (independent of species, but varying in space and time), introduced to satisfy the mass conservation, and evaluated as:

$$\mathbf{V}_C = -\sum_{k=1}^{NC}\mathbf{V}_k \tag{8.50}$$

The mixture diffusion coefficient $\Gamma_{k,mix}$ for species k-th is calculated using the following expression [11]:

$$\Gamma_{k,mix} = \frac{\sum_{j\neq k}^{N_S} X_j W_j}{W_{mix}\sum_{j\neq k}^{N_S}\frac{X_j}{\Gamma_{jk}}} \tag{8.51}$$

where W_{mix} and W_j are the molecular weights of mixture and species j-th, respectively. Γ_{jk} is the binary mass diffusion coefficient between species j-th and k-th. The remaining mixture-averaged transport properties are estimated from the corresponding pure-species properties through the application of proper mixing rules. In case of dynamic viscosity, the Wilke formula [12] is adopted:

$$\mu = \sum_{i=1}^{N_S}\frac{X_i\mu_i}{\sum_{j=1}^{N_S} X_j\phi_{i,j}} \tag{8.52}$$

where:

$$\phi_{i,j} = \frac{1}{\sqrt{8}}\sqrt{\frac{W_j}{W_i+W_j}}\left[1+\sqrt{\frac{\mu_i}{\mu_j}}\left(\frac{W_j}{W_i}\right)^{1/4}\right]^2 \tag{8.53}$$

For the thermal conductivity the combination averaging formula proposed by Mathur et al. [13] is used:

$$\lambda = \frac{1}{2}\left[\sum_{i=1}^{N_S} X_i\lambda_i + \left(\sum_{i=1}^{N_S}\frac{X_i}{\lambda_i}\right)^{-1}\right] \tag{8.54}$$

8.3.2.2 Boundary Conditions

The conservation equations of total mass, momentum, individual species mass fractions, and energy require boundary conditions to be specified for pressure, velocity, species mass fractions, and temperature. The usual boundary conditions for pressure and velocity are imposed. In particular, the pressure value is fixed at the outlet boundaries and a zero-gradient condition is imposed at the inlets and at the walls. For the velocity field, no-slip conditions are assumed at the walls. At the inlet, the velocity profile is assigned, whereas at the outlet boundaries the flow is assumed to be fully developed and zero-gradient conditions are then imposed.

Boundary conditions for the gas-phase-species mass fractions and temperature are summarized in the following.

Inert walls: The mass flux of the individual species k-th is set equal to zero:

$$\nabla\omega_k\big|_{inert} = 0 \tag{8.55}$$

According to the physics of the particular problem under investigation, on inert walls the temperature can be prescribed through a generic function of time $T_{inert}(t)$ or calculated to take into account the heat transfer with the external environment (at temperature $T_{env}(t)$):

$$T\big|_{inert} = T_{inert}(t) \tag{8.56}$$

$$(\lambda\nabla T)\big|_{inert} = U\left[T\big|_{inert} - T_{env}(t)\right] \tag{8.57}$$

where U is the global heat-transfer coefficient.

Catalytic walls: The mass flux of the individual species k-th is assumed equal to the formation rate due to the heterogeneous reaction occurring on the catalytic wall:

$$\rho\Gamma_{k,mix}(\nabla\omega_k)\big|_{catalytic} = \frac{A_{cat}^{eff}}{A}\dot{\Omega}_k^{het} \qquad k = 1,\ldots,NCG \tag{8.58}$$

Analogously, the heat flux is set equal to the heat Q^{het} released by the heterogeneous reactions occurring on the catalytic wall:

$$(\lambda \nabla T)|_{catalytic} = \frac{A_{cat}^{eff}}{A} Q^{het} \tag{8.59}$$

In the expressions reported above A_{cat}^{eff} is the effective catalytic area and A is the geometric area [14].

Inlet boundaries: At inlet boundaries, Danckwerts' conditions are set for gas-phase species, i.e. the total mass flux for each species k-th (accounting for diffusion and convection) is specified. If composition gradients exist at the boundary, these conditions allow diffusion into the computational domain, therefore giving a more accurate description than classic Dirichlet conditions:

$$(\rho \mathbf{v} \omega_k)|_{inlet} - (\rho \Gamma_k \nabla \omega_k)|_{inlet} = (\rho \mathbf{v} \omega_k)_0 \tag{8.60}$$

where the term on the right side is the prescribed total mass flux. In analogy, for temperature:

$$\left(\rho \mathbf{v} \hat{H}\right)\Big|_{inlet} - (\lambda \nabla T)|_{inlet} = \left(\rho \mathbf{v} \hat{H}\right)_0 \tag{8.61}$$

Outlet boundaries: At the outlet boundaries, the mass flux of gas-phase species and the heat flux are assumed equal to zero:

$$\nabla \omega_k|_{inert} = 0 \tag{8.62}$$

$$\nabla T|_{outlet} = 0 \tag{8.63}$$

8.3.2.3 Adsorbed (Surface) Species and Evaluation of the Reaction Term

In order to determine the formation rate due to the heterogeneous reactions occurring on the catalyst surface (8.58), a conservation equation which balances the rate of change of each surface species with the net formation rate due to heterogeneous chemical reactions at the surface has to be accounted for. In this scope, two different approaches can be adopted: (i) the mean-field approximation and (ii) kinetic Monte Carlo simulations.

8.3.3 Coupling CFD and Mean-Field Microkinetic Modeling

The mean-field approximation assumes that due to the strong mobility of the intermediates at the catalyst surface, the distribution of adsorbates and active sites is uniform. Thus, the local state of the active surface is well described by mean values of species coverage. Thus, the reaction rate can be locally expressed as a function of the mean surface molar concentration:

$$r_i = k_i \prod_{j=1}^{NR_{het}} X_{sur,j}^{\nu_{i,j}} \tag{8.64}$$

where k_i is the rate constant of the i-th step, $X_{sur,j}$ are the surface species mole concentrations, and $\nu_{i,j}$ is the stoichiometric coefficient of the j-th species in the i-th reaction [15]. Then, the evolution of the surface coverage is described by:

$$\sigma_{cat} \frac{\partial \theta_i}{\partial t} = \dot{\Omega}_i^{het} \qquad i = 1, \ldots, NCS \tag{8.65}$$

where θ_i is the site fraction of the species i-th, σ_{cat} is the site density, and $\dot{\Omega}_i^{het}$ the formation rate of species i-th due to the heterogeneous reactions. NCS is the number of absorbed surface species.

The use of detailed microkinetic models in the framework of CFD simulations poses important issues from a numerical point of view and requires specific algorithms. In particular, the classical fully segregated approaches usually employed in most of the non-reactive CFD simulations are not applicable because of the combination of the stiffness of the elementary reactions with the typical large dimensions of a microkinetic model [16]. Two alternatives are then possible. The first possibility is to employ fully coupled methods [17], in which all the processes are considered simultaneously, thus all the physical interactions among them are taken into account together. However, due to the prohibitive computational cost, these are strongly limited for complex geometries and for extremely large governing equations (especially when detailed kinetics are considered). The alternative solution is the application of operator-splitting algorithms, which allow the best numerical method to be used for each operator of the equations (typically reaction and transport). The resulting algorithms can be very complex and usually differ from term to term. An example of this implementation is given by Maestri and Cuoci [16], where the operator-splitting algorithm is implemented within the OpenFOAM framework [16, 18].

8.3.4 Coupling CFD and Kinetic Monte Carlo Simulations

When statistical simulations are used to explicitly account for the interplay between all the chemical events, master-equation-based *kinetic Monte Carlo* algorithms are employed in order to account directly for the spatial arrangements of the species at the surfaces and allow for a sound evaluation of the reaction term. The inclusion of kinetic Monte Carlo simulations in the framework of real reactor models requires dealing with very different time scales between the macroscale and mesoscale. In fact, the calculation of the reaction rate at each time step and in each part of the domain needs the explicit run of several kMC simulations. Since the characteristic time for kMC simulations is an order of magnitude lower than the one of the macroscale, this limits dramatically the resolution of the time step of the macroscale, thus leading to unbearable computational costs [7, 19–21]. Key in this respect is the development of methodologies to minimize the number of kMC simulations that have to be run "on the fly" during the simulation of the macroscale. The two main important strategies reported to solve this problem are reported below.

1. *Instantaneous steady state approximation*

This approximation exploits the typically different time scales of surface kinetics and gas flow. It is assumed that upon a change of the local gas-phase conditions (temperature, pressure, species mass fractions), the surface adapts instantaneously to the corresponding steady state. Thus, it is sufficient to have a continuous representation of the steady-state reaction rates as a function of the gas-phase conditions (temperature, pressure, species mass fractions) for coupling the kMC with the macroscale models. Due to the high computational time required for the kMC simulations in different conditions, the steady-state reaction rates can be pre-computed once before the whole reactor simulation is performed and be used to determine a numerically efficient continuous representation either using tabulation techniques or suitable polynomial fitting. Then, during the simulation, the species formation and consumption rates are evaluated by making use of the stored results [19–22].

2. *Gap-tooth methods*

In this approach, the kMC simulations are run only in some points of the surface domain and for the other points the reaction rates are interpolated. The converged solution is reached with iterative simulation of the whole domain [23–32]. This approach is used by Schaefer and Jansen [33] to couple the kMC simulations to a 1D finite-difference model for the case of CO oxidation. This scheme is used to reduce the number of kMC simulations carried out to estimate the reaction rates. In particular, due to the choice of describing the macroscale with a 1D finite-difference model, Schaefer and Jansen [33] select some preferential points where the kMC simulations are done and they estimate the reaction rates between these points with a linear interpolation. The correct description of the reactor behavior can be achieved only if the solution of the whole reactor simulations does not depend on the position of the points where the kMC simulations are done. On the contrary, Broadbelt and co-workers [34–36] use a methodology that updates the concentration and the reaction rate at the same time in all the discretization points. The main advantage of this approach is the possibility of describing also non-linear profiles between the points where the kMC simulations are carried out.

8.4 Case Study: Detailed Analysis of a Spectroscopic Cell Reactor

In order to exemplify the concepts reported in the previous sections, here we provide as showcase a detailed multiscale analysis of a spectroscopic chamber reactor. In particular, we apply the mean-field approximation coupled with CFD to study a real reactor geometry used for in situ spectroscopic studies [37]. Therefore, the structure of the cell is expected to play a crucial role in the coupling between chemistry and transport. The interactions between these two aspects generally modify the observable catalytic function in a nontrivial manner, as outlined in the previous sections.

8.4.1 Reactor Geometry and Operating Conditions

The design of the chamber is heavily constrained by the invasive machinery and this results in very complex flow profiles characterized by temporal and spatial inhomogeneity, as we exemplify by analyzing the prototypical geometry shown in the left picture of Fig. 8.6. Without considering any specific invasive instruments or pumping equipment, the geometry consists of a cube of edge length 3 cm, within a cuboid sample holder with a base of 1 $\times$ 1 cm and a height of 1.5 cm mounted on the bottom wall. The catalyst surface covers the entire top face of the cuboid. The gas flows into the reactor through an inlet located at the upper left and exits through an outlet placed at the lower right. External heating keeps a constant temperature at both the catalyst surface and the sample holder, whereas the remaining walls are considered adiabatic. For the CFD simulations, we exploited the mirror symmetry of the geometry, thus one-half of the chamber has been discretized with a regular cubic mesh. The resolution is 0.5 mm and it is tightened to 0.25 mm in the proximity of the catalyst surface to adequately describe the concentration and temperature gradients, as shown in the right panel of Fig. 8.6. This results in 224,000 computational cells. Simulations were performed using the catalyticFoam solver [16] in the OpenFOAM environment [18].

We focused on methane steam reforming on Rh employing the detailed microkinetic mechanism by Maestri and co-workers and Vlachos and co-workers [38, 39], which consists of 21 gas species and 13 adsorbed species involved in 82 surface reactions. The gas mixture (97% N_2, 2% H_2O, 1% CH_4, in molar basis) enters at the inlet section and a flowrate of 120 Nm L/min through the reactor, which falls within the common range characteristic for such applications. The reactor works at atmospheric pressure.

As initial conditions, we adopted uniform values for all fields in the reactor, employing the inlet value for temperature, the outlet value for pressure. The reactor was assumed to be full of nitrogen at the starting time. The solution is advanced

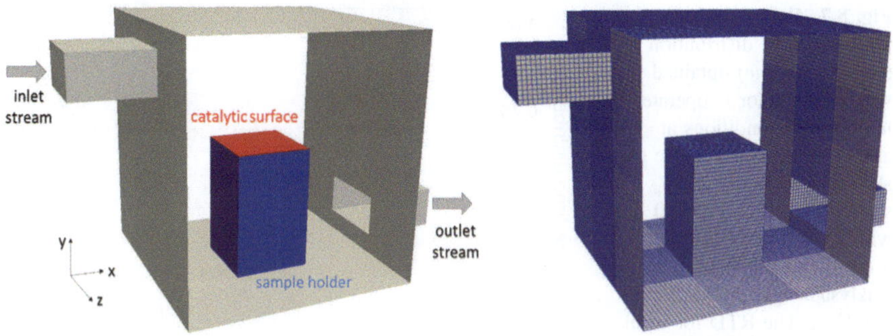

Fig. 8.6 Computational domain (on the *left*) and mesh (on the *right*) adopted for the CFD simulations. The gas mixture enters with a uniform velocity and fixed temperature from the *left side*, streams over the catalytic surface (in *red*), leaving the domain through the outlet section on the *right side*. The temperatures of the catalytic plate and the sample holder are kept fixed. All remaining walls can be kept at the same temperature of catalytic plate and sample holder (isothermal simulations) or can be considered perfectly adiabatic (adiabatic simulations). Because of the symmetry of the system only one-half of the computational domain is meshed. Cells with two different sizes were adopted: 0.25 and 0.50 mm. The resulting, total number of computational cells is equal to 224,000

using a Courant number of 0.05 until the simulation reaches a steady state in which the relevant flow features are fully established, as proposed in [16, 40].

The gas enters the reactor at the inlet with a temperature of 300 K and both the catalyst and the sample holder are kept at a fixed temperature. In particular, the effect of different catalyst temperatures is investigated, assuming a catalyst/sample holder temperature of 973 and 673 K.

8.4.2 Reactor Analysis: Residence Time Distribution

To investigate the effect of both flow and mixing on the behavior of the reactor, the residence time distribution (RTD) has been analyzed. The nominal residence time for the reactor in the actual operating conditions, i.e. the ratio between the volumetric inlet flow rate and the volume of the chamber, is equal to 12.86 s. We performed an RTD analysis of the prototypical spectroscopic reactor to investigate the effect of the complex flow field on the distribution of residence time. Figure 8.7 shows the RTD for the real reactor and for the two ideal limits described in Sect. 8.2. The RTD profile presents a peak at low residence time showing the presence of a bypass, which means that a preferential path exists, followed by the gas to travel from the inlet to the outlet. Indeed, the gas enters and drops down due to density, reaching the bottom of the chamber and rapidly streams out of the reactor. The tail of the profile shows the presence of quiescent zones, where the elements of fluid last for a time larger than the nominal residence time. In the actual operative conditions, the calculated mean residence time is 4.88 s and 6.46 s for the

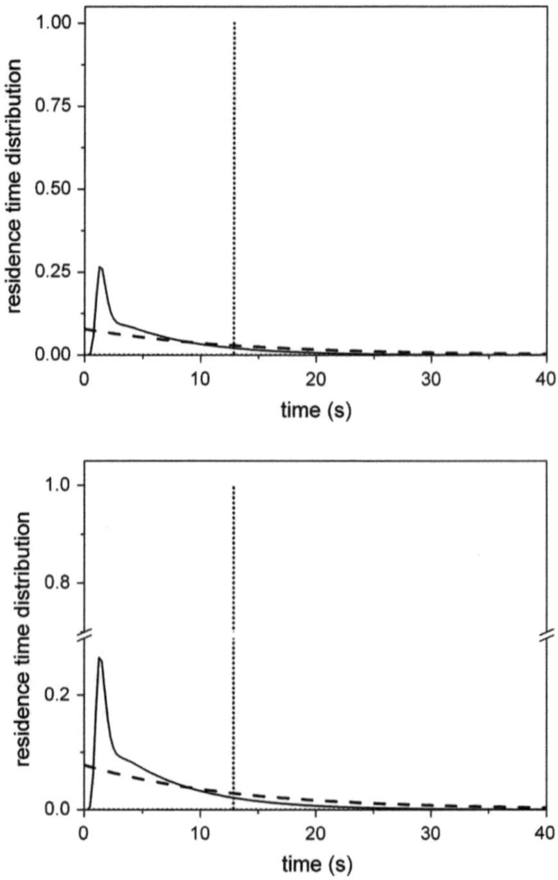

Fig. 8.7 Calculated residence time distribution (*continuous line*) obtained when the reactor is operated in adiabatic conditions at atmospheric pressure. The gas stream through the inlet section takes place at 300 K, with a uniform velocity of 4 cm/s, and the catalyst/sample holder is kept at 973 K. The RTD for ideal behavior of plug flow (*dotted line*) and perfect mixing (*dashed line*) are plotted for comparison

catalyst/sample holder maintained at 973 K and 673 K, respectively. Both values are significantly lower than the nominal value. In this view, the approximations of perfect mixing and of plug flow are not able to describe the behavior of the reactor, requiring a more accurate analysis of the flow pattern inside the chamber.

The non-idealities predicted accordingly to the RTD analysis are highlighted by the streamlines shown in Fig. 8.8. The denser gas streams into the reaction chamber from the cold inlet, drop down reaching the bottom of the reactor, and flow towards the outlet. This preferential path is followed by most of the fluid, resulting in the peak of the RTD profile. A small portion, after reaching the hot sample holder, heats up and, due to buoyancy, rises inducing a vortex, which rotates counter-clockwise around the catalyst. This phenomenon increases the residence time of some marked elements of gas, leading to the long tail representative of dead zones. Therefore, the flow field inside the reactor is strongly influenced by a significant degree of thermo-convection, raised in the pseudo-stationary state, leading to a highly complex velocity pattern.

8.4.3 Reactor Analysis: Species Concentration and Temperature Fields

The complex flow field within the chamber, especially close to the sample holder, influences the concentration of the reactants in proximity of the catalyst surface. Figure 8.9a shows the predicted CH_4 mass fraction distribution within the catalyst chamber obtained with the catalyst/sample holder at 673 K. Strong non-homogeneities in the spatial distribution of methane experiences the important effect of the interactions between transport and chemistry. For the higher sample holder temperature, the catalyst ends up in a highly reactive state, leading to strong mass-transfer limitations, as apparent from the calculated mass fraction of CH_4 inside the reactor. The controlling phenomenon is the species diffusion from the bulk of the gas phase to the catalyst surface, since the reaction rates are extremely fast. The heat conduction and convection in the reactor are not sufficiently strong to make the temperature uniform, with the exception of a small region close to the catalyst/sample holder. When the cold inlet gas enters the chamber, it drops down due to the difference of density, without having enough time to heat up. Thus, the region of the reactor below the inlet section shows a temperature of ~250 K lower with respect to the sample holder temperature. The species and temperature concentration inside the chamber is non-uniform, and strong gradients exist between the bulk and the catalyst surface, due to the combined effect of complex flow field along with high reactivity of the catalyst determining external mass-transfer control. The gas-phase composition in the proximity of the catalyst strongly relies on transport phenomena, making the assumption of perfect mixing within the chamber unfeasible. On the other hand, the hypothesis of plug flow is incorrect due to the recirculation induced by the difference of density, which produces an additional

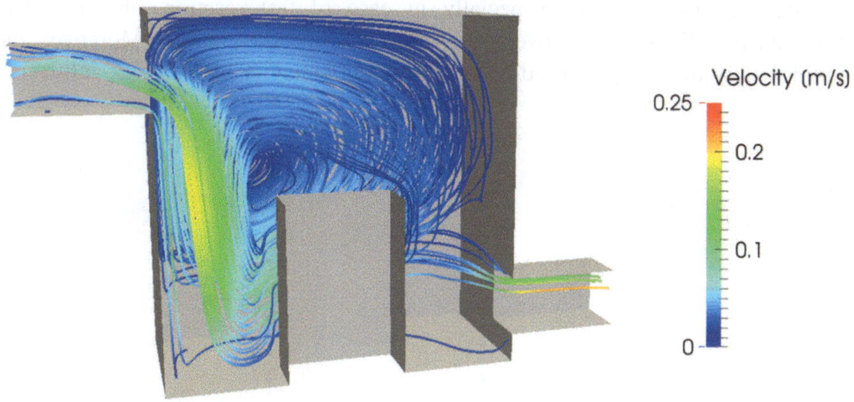

Fig. 8.8 Streamlines arising when the reactor is operated in adiabatic conditions at atmospheric pressure. The gas stream (1% CH_4, 2% O_2, 97% N_2, on a molar basis) enters at the inlet section at 300 K, with uniform velocity of 4 cm/s, the catalyst/sample holder is maintained at 973 K. The streamlines are colored using the velocity field magnitude

mixing. Nevertheless, even if the catalyst should operate under less reacting conditions (e.g. by decreasing the sample holder temperature), the complex flow field would render it impossible to assume a uniform distribution of species and temperature. Due to the lower reaction rates, the characteristic time of the diffusion process from the gas phase to the surface becomes at most of the same order of magnitude of the reaction rates. The controlling process is the rate of the heterogeneous chemistry and, even in this context, a concentration gradient rises from the surface to the bulk of the gas phase. The predicted temperature distribution, reported in Fig. 8.9b, highlights the presence of a colder region in the left side of the chamber due to the cold inlet stream. Once again, the predicted distributions of species and temperature are inhomogeneous. In particular, the concentration near the catalyst surface is different from the outlet concentration, making the information recovered from the outlet mass-flow rate not totally representative of the state of the reactor.

The complex flow field induced by the interactions between the reactor geometry, the transport phenomena, and the surface reactivity alters the species distribution on the catalyst surface. The calculated mass fractions along the main axes, presented in Fig. 8.10a, b, highlight the presence of spatial gradients in both directions. In particular, the recirculation moves the gas toward the left side of the catalyst, altering the transport rate of the species to the catalytic surface. Overall, the species concentrations are non-uniform, pointing out the importance of taking into account the macro-scale phenomena to obtain an accurate description of the reactor. In particular, an analysis by means of CFD allows for the clarification of the coupling between macroscale transport processes and surface chemistry, leading to a better understanding and design of the experiments. The non-uniform gas composition on the catalyst surface affects the site-coverage distribution, which shows the same trends of the gas phase species, as shown in Fig. 8.10c, d.

The interaction between transport and chemistry strongly affects the behavior of the spectroscopic chamber, especially in case of fast reactions that establish mass-transport limitations between the gas phase and the catalyst. Moreover, the complex flow field, experienced by the non-ideal behavior of the RTD, influences

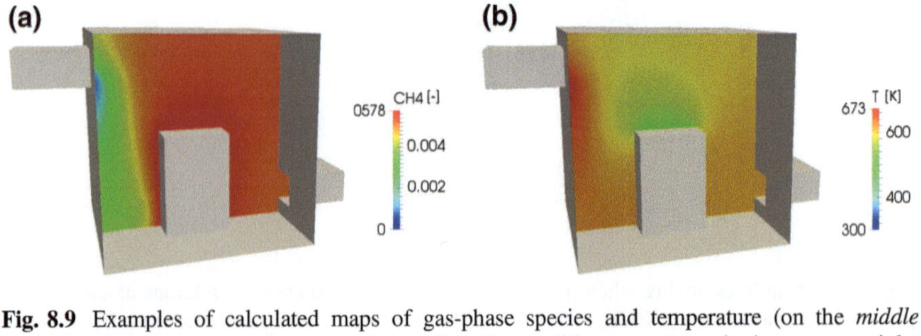

Fig. 8.9 Examples of calculated maps of gas-phase species and temperature (on the *middle* section) when the reactor is operated in adiabatic conditions at atmospheric pressure, inlet temperature of 300 K, and sample holder/catalyst temperature of 673 K

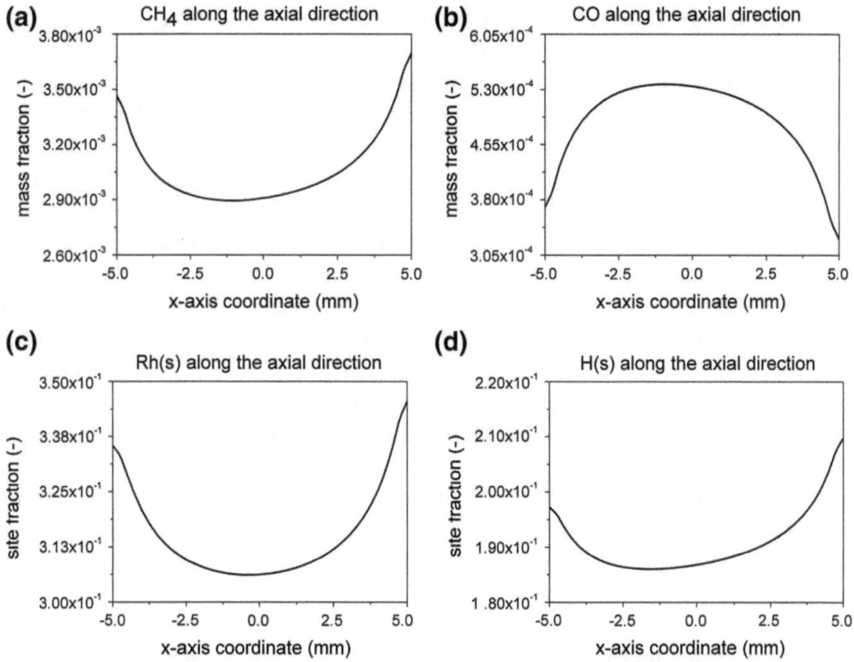

Fig. 8.10 Examples of calculated mass fraction profiles of selected gas-phase species along the axial direction of catalytic plate (x coordinate) for catalyst sample holder at 673 K, *panels* (**a**) and (**b**). Example of calculated site coverage profiles of selected adsorbed species along the axial direction of catalytic plate (x coordinate) for catalyst sample holder at 673 K, *panels* (**c**) and (**d**)

the distribution of species across the reactor, leading to non-homogeneities, which have to be accounted for in order to derive a detailed understanding of the catalyst material in reacting conditions. These effects have to be considered when aiming to relate *operando* spectroscopic data obtained in the cell with the activity of the catalyst. In this context, CFD is a useful tool to obtain a deep insight into the behavior of the spectroscopic reactor, with the aim of understanding the real interactions between chemical and transport processes.

8.5 Summary and Outlook

In heterogeneous catalysis 'active sites' are, by nature, present in a separate solid phase. Thus, the reactants in the gas phase have to reach the 'active sites' and this involves several physical transport processes. All these steps occur at a finite rate and consequently concentration and temperature gradients may arise between the bulk fluid phase and the areas in proximity of the active site. Therefore, the physical steps may have a strong impact on the rate of the overall process and the observable reaction rate may differ substantially from the intrinsic reaction rate of the chemical transformation under bulk fluid-phase conditions.

Thus, it is of crucial importance to distinguish between the intrinsic functionality of the catalyst material and its observed macroscopical behavior in the reactor, i.e. the device where the physicochemical transformations occur. By reviewing basic concepts of chemical reaction engineering, we have shown the multiscale character of the observed macroscopic behavior of the catalytic process. The modes of operation of the chemical reactor strongly influence the performance of a given material, both in terms of hydrodynamics (and its immediate effect on the residence time distribution) and in terms of transport properties. Therefore, a thorough understanding of chemical reaction engineering is a crucial requirement in order to properly design reaction equipment or to correctly interpret the experimental evidence. In this view, fundamental multiscale modeling is appearing in the last decade as one of the main frontiers in chemical reaction engineering, especially contributing to the quest of advancing our scientific understanding of multiscale kinetic transport interactions to enable better reactor choice and to ensure higher reactor and process efficiencies. Specifically, the main task is to provide an improved scientific basis for designing the chemical transformation in a seamless flow of information from the atoms to the reactor.

From a reaction engineering perspective, CFD simulations have matured into a powerful tool for understanding mass and heat transport in catalytic reactors, by coupling the flow field and heat-transport models with models for heterogeneous chemical reactions. The most natural approach would be to proceed according to a bottom-up philosophy, from the atomic scale to the reactor scale. However, this approach is hampered by several issues, which mainly concern with the coupling of very different time scales. Indeed, the resolution of the macroscopic scale with the small time step required for the microscale would lead to an enormous effort in computational time since the continuum dynamics usually evolves on a much larger time scale. In this respect, we have shown that most of the methods proposed and applied in the literature for the simulation of gas-solid catalytic reactors are based on an effective decoupling of the interdependencies among the scales. The main and general feature of these methods is to treat separately each scale either by using specific numerical procedures or by introducing specific approximations in order to effectively decouple the time scales of the different scales. In doing so, each scale is simulated with different time steps, thus avoiding the direct coupling of phenomena occurring at very different characteristic times. Such methods enable the incorporation of both complex mean field microkinetic models and kinetic Monte Carlo Simulations in the fundamental modeling of real reactor geometries. We have exemplified the applications of these methods through the CFD analysis of a spectroscopic cell. This study has clearly shown that transport artifacts may arise due to the complicated geometries usually employed, and that this can lead to an erroneous interpretation of the experiments, if not properly accounted for.

As a whole, fundamental multiscale chemical reaction engineering represents one of the most important progress areas for catalysis engineering. By bridging between chemistry and engineering science, physics and material science, it draws from hitherto rather disjoint communities and therefore it is ideally suited for broad interdisciplinary collaborations with research groups with specific and complementary expertise (e.g. material synthesis, *operando* and in situ characterization, kinetic experiments, transport phenomena and chemical engineering).

References

1. O. Deutschmann, *Modeling and Simulation of Heterogeneous Catalytic Reactions: From the Molecular Process to the Technical System* (Wiley, 2013)
2. M.P. Dudukovic, Science **325**, 698 (2009)
3. M.K. Sabbe, M.-F. Reyniers, K. Reuter, Catal. Sci. Technol. **2**, 2010 (2012)
4. M. Maestri, K. Reuter, Angew. Chem. Int. Ed. **50**, 1194 (2011)
5. M. Salciccioli, M. Stamatakis, S. Caratzoulas, D.G. Vlachos, Chem. Eng. Sci. **66**, 4319 (2011)
6. G.F. Froment, K.B. Bischoff, J. De Wilde, *Chemical Reactor Analysis and Design*, vol. 2 (Wiley, New York, 1990)
7. K. Reuter, D. Frenkel, M. Scheffler, arXiv:cond-mat/0408080 (2004)
8. M.R. Charest, C.P. Groth, Ö.L. Gülder, Combust. Theor. Model. **14**, 793 (2010)
9. S. Chapman, T.G. Cowling, *The Mathematical Theory of Non-uniform Gases: An Account of the Kinetic Theory of Viscosity, Thermal Conduction and Diffusion in Gases* (Cambridge University Press, 1970)
10. T. Coffee, J. Heimerl, Combust. Flame **43**, 273 (1981)
11. R.B. Bird, W.E. Stewart, E.N. Lightfoot, *Transport Phenomena*, 2nd edn. (Wiley, New York, 2002)
12. C.R. Wilke, J. Chem. Phys. **18**, 517 (1950)
13. S. Mathur, P.K. Tondo, S.C. Saxena, Mol. Phys. **12**, 569 (1967)
14. M. Maestri, A. Beretta, T. Faravelli, G. Groppi, E. Tronconi, D.G. Vlachos, Chem. Eng. Sci. **63**, 2657 (2008)
15. M. Maestri, D.G. Vlachos, A. Beretta, G. Groppi, E. Tronconi, A.I.Ch.E. J. **55**, 993 (2009)
16. M. Maestri, A. Cuoci, Chem. Eng. Sci. **96**, 106 (2013)
17. M.D. Smooke, R.E. Mitchell, D.E. Keyes, Combust. Sci. Technol. **67**, 85 (1986)
18. H. Jasak, A. Jemcov, Z. Tukovic, *International Workshop on Coupled Methods in Numerical Dynamics* (2007), p. 1
19. S. Matera, K. Reuter, Catal. Lett. **133**, 156 (2009)
20. S. Matera, K. Reuter, arXiv:1006.0343 (2010)
21. S. Matera, K. Reuter, Phys. Rev. B **82**, 085446 (2010)
22. S. Matera, M. Maestri, A. Cuoci, K. Reuter, ACS Catal. **4**, 4081 (2014)
23. A. Armaou, I.G. Kevrekidis, C. Theodoropoulos, Comput. Chem. Eng. **29**, 731 (2005)
24. A. Bindal, M.G. Ierapetritou, S. Balakrishnan, A. Armaou, A.G. Makeev, I.G. Kevrekidis, Chem. Eng. Sci. **61**, 779 (2006)
25. C. Gear, I.G. Kevrekidis, arXiv:physics/0211043 (2002)
26. C.W. Gear, J. Li, I.G. Kevrekidis, Phys. Lett. A **316**, 190 (2003)
27. W. Gear, I. Kevrekidis, J. Hyman, P. Kevrekidis, O. Runborg, C. Theodoropoulos, Commun. Math. Sci. (2003)
28. I.G. Kevrekidis, C.W. Gear, G. Hummer, AlChE J. **50**, 1346 (2004)
29. I.G. Kevrekidis, C.W. Gear, J.M. Hyman, P.G. Kevrekidid, O. Runborg, C. Theodoropoulos, Commun. Math. Sci. **1**, 715 (2003)
30. A. Roberts, I. Kevrekidis, ANZIAM J. **46**, 637 (2005)
31. G. Samaey, I.G. Kevrekidis, D. Roose, *Multiscale Modelling and Simulation* (Springer, 2004), p. 93

32. G. Samaey, D. Roose, I.G. Kevrekidis, Multiscale Model. Simul. **4**, 278 (2005)
33. C. Schaefer, A. Jansen, J. Chem. Phys. **138**, 054102 (2013)
34. L.J. Broadbelt, R.Q. Snurr, Appl. Catal. A **200**, 23 (2000)
35. D.J. Dooling, L.J. Broadbelt, Ind. Eng. Chem. Res. **40**, 522 (2001)
36. D. Majumder, L.J. Broadbelt, AlChE J. **52**, 4214 (2006)
37. A. Stierle, A.M. Molenbroek, MRS Bull. **32**, 1001 (2007)
38. M. Maestri, D. Livio, A. Beretta, G. Groppi, Ind. Eng. Chem. Res. **53**, 10914 (2014)
39. M. Maestri, D.G. Vlachos, A. Beretta, G. Groppi, E. Tronconi, AlChE J. **55**, 993 (2009)
40. T. Maffei, S. Rebughini, G. Gentile, S. Lipp, A. Cuoci, M. Maestri, Chem. Ing. Tech. **86**, 1099 (2014)

Index

J. Frenken and I. Groot (eds.), Operando *Research in Heterogeneous Catalysis*, Springer Series in Chemical Physics 114, DOI 10.1007/978-3-319-44439-0

MIX
Papier aus verantwortungsvollen Quellen
Paper from responsible sources
FSC® C105338

If you have any concerns about our products, you can contact us on
ProductSafety@springernature.com

In case Publisher is established outside the EU, the EU authorized representative is:
Springer Nature Customer Service Center GmbH
Europaplatz 3, 69115 Heidelberg, Germany

Printed by Libri Plureos GmbH
in Hamburg, Germany